Instructor's Manual for

Statistics

Second Edition

Instructor's Manual for

Statistics

Second Edition

David Freedman
Robert Pisani
Roger Purves
Ani Adhikari

W. W. Norton & Company
New York and London

Cartoons by Dana Fradon

Printed in the United States of America.

Library of Congress Cataloging in Publication Data
Instructor's manual for Statistics, 2nd edition / David Freedman...
[et al.].
p. cm.
1. Mathematical statistics. I. Freedman, David, 1938–
II. Freedman, David, 1938– Statistics.
QA276.I58 1991
519.5—dc20 90-49480

ISBN 0-393-96044-7

W. W. Norton & Company, Inc., 500 Fifth Avenue, New York, N.Y. 10110
W. W. Norton & Company, Ltd., 10 Coptic Street, London WC1A 1PU

1 2 3 4 5 6 7 8 9 0

CONTENTS

INTRODUCTION TO THE SECOND EDITION

After the first edition of *Statistics* was in print for 10 years, it began to show definite signs of age; for example, by 1987, the 1977 prices and incomes were out of date by a factor of 2. More troubling, some of the main data sets were collected well before the class of 1987 was born. We decided to revise the book a little, putting in new data and fixing some of the passages we liked least. However, one thing led to another, and two years later we finished by doing a major rewrite.

The topics, and the general order of presentation, remain about the same; so does the flavor of the text. We tried to make the exposition tighter, with fewer repetitions of material. This made room for many new examples and exercises. Chapter 2 on observational studies and chapter 9 on correlation were completely rewritten, and are much more focused. Chapters 13 and 14 on probability were reorganized; conditional probabilities are now discussed, and nonstandard terminology was eliminated. Chapter 27 on the two-sample z-test was expanded, to cover randomized trials.

The second edition has been class-tested at Berkeley, Stanford, and Utah State. We think it has all the same strengths as the first edition, and fewer weaknesses.

INTRODUCTION TO THE FIRST EDITION

The world is full of elementary statistics books. Why are we publishing another one? The answer is that we came to want a book which would explain the basic ideas in the subject to an intelligent but nonmathematical reader, and make the ideas vivid through real examples. These objectives seem innocent enough, but writing a book to meet them turned out to be much harder than we had expected. We proceeded largely by trial and error, going through six drafts before publication. Each draft was tested for a year on many hundreds of Berkeley undergraduates, in courses at different levels of difficulty (with or without a calculus prerequisite), the class sizes ranging from 10 to 300.[1] Each year, we watched the students working on the materials, listened carefully as the friendlier ones told us what was wrong with the exposition, and scribbled frantically away at the next year's draft.

---

[1]The book was mainly developed in the Statistics 2 course at Berkeley. This course, which is divided into two or three large lecture sections, enrolls about 500 students a quarter, drawn mainly from business administration, the social sciences, and the less quantitative natural sciences. Still, about 40% of these students have had a college calculus course, and about 20% of them have completed two or more additional college mathematics courses.

The book has also been used extensively in Statistics 20, with class sizes of 30; virtually all students in this course have had calculus, and about half are in quantitative fields like mathematics, statistics, computer science, and engineering.

Along the way, we were forced to notice some unpleasant facts. The first shock was discovering how much trouble the students had with arithmetic. In self-defense, we started giving pre-tests. The one from fall, 1977 is reproduced at the end of this manual: similar tests have been given to several thousand students.

Here are four questions from the fall, 1977, pre-test.

1. 300 is what percent of 2,000?

5. In the United States, 1 person out of every 500 is in the navy and one-sixth of naval personnel are officers. What fraction of the United States population consists of naval officers? Or can this be determined from the information given?

6. $\sqrt{100{,}000}$ is about:
(i) 30 (ii) 300 (iii) 1,000 (iv) 3,000 (v) can't tell

11. A quart of vodka is 40% alcohol. Write a formula for the percentage of alcohol in a mixture of V quarts of vodka and J quarts of orange juice.

Only three Berkeley statistics students in four can do the percentage in question 1, and only two in three can handle the square root in question 6. Question 5 tests whether they know when to multiply fractions; only one student in four gets it right. Many elementary statistics texts claim their sole prerequisite to be "high school algebra." Question 11 is a very gentle probe into what the students remember from high school algebra: one student in six can write down the formula.[1]

[1] The pretest from fall 1988 is also reproduced below. By then, Berkeley claimed to have tightened up its admissions standards; but the students seemed to know less.

The pre-test even seems to understate matters. One issue it misses is reliability: a student may be quite good at doing one-line arithmetic problems (like, $\sqrt{2,500}$ = _____); but may not be able to handle an exercise which requires doing half a dozen steps of similar difficulty. Another issue is context: when students have trouble deciding which arithmetic operations to perform in response to word problems, many stop being able to do arithmetic at all. It is as if they get exhausted during the analysis phase.

Now when we started writing, we tried to teach the conventional notation: $\sum_{i=1}^{n}(x_i - \bar{x})^2$, and all the rest. But it soon became clear that the algebra was getting in the way. For students with limited technical ability, mastering the notation demands so much effort that nothing is left over for the ideas. To make the point by analogy, it is as if most the undergraduates on the campus were required to take a course in Chinese history--and the history department insisted on teaching them in Chinese.

So we decided to try writing in ordinary English. For three probabilists, this presented some unexpected difficulties. And it led to a surprise in the classroom: the students wanted the equations, even though they found the symbolism baffling. Perhaps we shouldn't have been surprised. Nonmathematical students seem to flounder in numbers courses. They survive only by ruthless pragmatism. Their objective is to pass the final. Usually, the final is a series of word problems, and the course is seen as a series of equations. The instructor may think that the equations express some general truths, but this tends to be lost on the students. For them, the main issue is learning how to associate the equations in the course with the word problems on the final, and recognizing which numbers in the word problem are to be substituted for which variables in the equation. There is a Berkeley student word for this syndrome: *pluginski*.

By the time students get to a statistics course, pluginski is so ingrained that anything like an equation tends to shortcut thought: students just grab the equation and run. Without equations, students really have to work at understanding the concepts in order to solve the problems. This is exactly what we want to achieve, even though the students find it irritating. As a result, whenever possible, we banish equations.[1] However, in many cases we do have a substitute: short summary sentences for the major points. There is a definite advantage to this approach: it is hard to memorize an English sentence without paying some attention to what the words mean.

By now, we had been through several drafts, and thought the worst was over. It wasn't. By our lights, we had succeeded in translating quite a lot of statistics into acceptable English. But, as we discovered, the students were still

---

[1]Some instructors who have used the book do not buy our hard-nosed attitude. They do the equations in lecture, and tell us the students accept this as complementing the text. There is a supplement which may be useful in this regard: *Mathematical Methods in Statistics*, by David Freedman and David Lane (W. W. Norton, New York, 1981).

having a hard time with our materials. We got discouraged enough to start grumbling to colleagues in other departments, showing them the "easy" passages the students couldn't read. The colleagues couldn't read them either. Where we saw simplicity, they saw a maze of argument.

This was a low point, but things improved from there. We realized that the problem wasn't "dumb students"; it was more a case of nonstatisticians seeing the world very differently from statisticians, needing different kinds of explanations, and wanting to learn different kinds of skills. Very few members of our audience are actually going to derive formulas, or carry out large scale data analysis. Many, however, are going to have to deal with statistical findings, because nowadays it is hard to read research journals--or even newspapers--without coming across statistical arguments.

We began to rethink our strategy. We had been making a tacit assumption: The exposition should start from the points which were clear and obvious--"elementary"--to us, building up to more complicated and interesting ideas. However, elementary mathematical points are often rather hard, even when expressed in English. Insisting on these points just confuses things, and distracts attention from the main issues. Also, we were still focusing on the procedures, leaving it to the students to infer the purposes of the activities: the scientific questions being answered.[1] This is fine for people who find technique easy, and therefore

[1] In retrospect, this seems a bit disingenuous; students are often much more interested in procedure than in purpose.

have time to think about what they're doing. For our readers, students and nonstatistical colleagues alike, this was a failure.

We decided to start at the other end: What are the main ideas that our field has to offer the intelligent outsider? Everything else, no matter what its technical interest, had to be set aside. Then, the reader has to be persuaded that each idea is worth knowing. To do that, we had to make explicit the question behind the statistical procedure. Often, we were able to find some vivid example embodying the question. Similarly, many statistical concepts formalize some understanding about the world; and in many cases we were able to find the right example to crystallize this insight. Once motivated, the ideas had to be presented in reasonably smooth language, free of annoying technicalities. And it all had to be fitted together into a coherent narrative, so that at each stage the reader would know enough to appreciate the next question.

Carrying out this program turned out to be a real adventure, because it forced us to reconsider the basics of the field from a different perspective. In the end, we think we brought it off. The book covers a good set of topics for a first course, arranged in logical order, and properly illustrated by examples. It works quite well for us, and for many friends who have tried it elsewhere.[1] Sample tests, with

[1]Minnesota, Sonoma State College, Stanford, UC Los Angeles, UC Santa Barbara, Wisconsin, Yale.

pass rates, are reproduced below.[1] As far as we can see, the book is intelligible to its intended audience: nonstatisticians who want to learn some statistics in order to go about their affairs. This includes students in college classrooms--as well as professionals in other fields.

To some statisticians, the book looks like an easy read--too easy to use as a college text. This criticism is off the mark. The book is intelligible, but not easy. We know from our courses that students, even those who have had a fair amount of college mathematics, have to work quite hard making their way through the book and solving the exercises. In part, this is because there are many pedagogical difficulties we just could not overcome. Then too, statistics does involve a lot of hard ideas. Instructors who use the book will have to help their students master these ideas.

[1]In typical Statistics 2 finals, the class averages were around 60 out of 100, with an SD of 20. Students with calculus averaged around 65, each additional college mathematics course contributing around 2 points to the average. On similar tests, Statistics 20 students, who know calculus and are majoring in quantitative fields, averaged about 70. Most of the test questions were taken from exercises in the book, so the students had seen them before.

An interesting sidelight: about 70% of the Statistics 2 students take the course to fulfill a requirement; the others take it voluntarily. Those taking it as a requirement only averaged about 55; the volunteers averaged over 65.

SCHEDULING

At Berkeley in the 1970s, Statistics 2 was taught in ten-week quarters, with three hours of lectures a week, and three hours of laboratory. In the 1980s, the university went back to fifteen-week semesters. The book can be used successfully with both calendars. It is written so that most chapters take about an hour of lecture time. However, this is a fairly quick pace. To maintain it, the more difficult sections in some chapters have to be skipped, or carried over to a second lecture.

There are 29 chapters to the book, so something has to go to fit it into a quarter. In a semester, the whole book can be covered; there may even be some time to spare at the end, to do some of the mathematical formalism. Dependencies among the various parts of the book have been minimised in the writing, leaving instructors fairly free to pick and choose.

As far as we are concerned, the logical core of the book consists of:

Chapters 3–4–5	Descriptive statistics
Chapter 13	What are the chances?
Chapters 16–17–18	Chance variability
Chapters 19–20–21, 23	Sampling

Sometimes when we teach the course, we cover parts I–VII, but omit part VIII on testing. At other times, we have covered everything except part III (correlation and regression). A third strategy which we can recommend is to cover the whole book, omitting

Chapter 12	The regression line
Chapter 15	The binomial coefficients
Chapter 25	Chance models in genetics
Section 26.6	The t-test
Sections 27.3–4	The z-test for experiments
Chapter 28	The chi-square test

EXERCISES

We discovered early on that unless we could write an exercise to test a point, students were not likely to learn it. So we worked quite hard to create a variety of good exercises. Most sections close with an exercise set, the answers being at the back of the book. All chapters but 1 and 7 include a set of "review exercises." In many chapters, the review exercises cover previous material too. This prevents the material from disappearing, and makes the students learn to judge when the different procedures apply. Answers to the review exercises do not appear in the book, but are given in the instructor's manual, below. We usually make out homework assignments from the review exercises, and put some of them on the tests as well. Generally, we assign about half the review exercises in the book as homework.

Most exercise sets include a few problems which can be solved by a straightforward application of the procedures just covered in the book. However, there usually are harder problems too. Some exercises, for instance, ask the students to decide whether a proposed procedure is sensible, or to choose among competing procedures. Other exercises ask the students to make rough guesses as to the magnitudes of certain quantities, still others call for qualitative judgments. Such exercises cannot be solved by mechanical application of formulas; they require understanding.

Many exercise sets can be used as diagnostic aids, to pinpoint the difficulties students are having with the concepts. We often get the students to do the exercises in laboratory periods, working with together in small groups. We go around from group to group, talking to them about what they are doing. The exercises provide a good framework within which to discuss the ideas we want to get across.

SUPERVISION

One key to teaching a large lecture course is supervision of the teaching assistants who handle section meetings (or "labs," in the Berkeley vernacular). Our experience is that TAs want to lecture: not unnaturally, they want to teach the mathematics they've just learned in their graduate courses, and are a little impatient with our nonmathematical approach. On the other hand, we think we've already given the lectures, and just want the TAs to help the students work problems.

To make this stick, we drop in on the labs from time to time, and observe the TAs at work, or talk to the students ourselves. More formally, we meet the TAs once a week, and review with them the problems to work in lab. This means going over the statistical content of the problems, and the pedagogical issues: what does this problem illustrate? where is it discussed in the text? what will students find hard? how can you break the problem down into smaller pieces? These sessions and the lab visits were eye-openers--for us and the teaching assistants alike.

At Berkeley, the students turn in homework; this is graded by "readers," often undergraduate majors. The readers work on a very tight time-table, and come out of a tradition where word problems have numerical answers which are right or wrong. On the other hand, we want solutions to be written out in reasonable style, with the logic explained. Grading should support these goals, but may in fact subvert them. The only way out is to spot-check the grading.

DETAILED COMMENTS ON THE SECOND EDITION

This section of the manual has detailed comments on the different chapters in the book, outlining the contents, pointing out any nonstandard language and some of the pedagogical difficulties. Many of the comments are directed to courses where the students have weak mathematical backgrounds. (This covers more territory than might be apparent.)

PART I. DESIGN OF EXPERIMENTS

The material in part I of the book is interesting and not very technical, so we find it a natural introduction to the subject. However, some instructors may wish to start right in with descriptive statistics (part II), and deal with design issues as they come up. The book is organized with this possibility in mind.

Chapter 1. Controlled Experiments

This chapter explains the key elements in a randomized controlled double-blind design, and why each is necessary. The context is the Salk vaccine field trial. Other examples are presented to reinforce the ideas.

Conventional wisdom dictates that the investigator should control the key variables and randomize the rest. The text focuses on the randomization, which is the hard idea. Some instructors will want to pay more attention to the possibility of controlling variables by stratifying subjects before randomizing.

Section 3 is new in the second edition.

Chapter 2. Observational Studies

In this chapter, observational studies are distinguished from controlled experiments. With an observational study, it is harder to draw conclusions about cause-and-effect relationships. The "cause" and "effect" may both be the result of some hidden third factor. (We return to this theme in chapter 9.) All the examples in this chapter are new in the second edition, except for the admissions study.

Notes on review exercises. Exercises 1 and 2 may seem unnecessary, but many students do not realize that you take percentages to adjust for differences in group sizes. Some such students think that with a bigger denominator, the percentage will be bigger; others, perhaps more sophisticated, think the reverse.

Exercise 12 covers Simpson's paradox (section 2.4). Exercises like number 9 are tough, because students don't see any alternative explanation to the causal one. In grading, we aren't sympathetic to rote repetition of slogans--even ones we believe, like "association isn't the same as causation."

Notes on lecturing. Instructors have asked us how we handle this part of the book in lecture, and we have done it several ways. One is to give a straightforward presentation of the material: each chapter can be covered in one lecture, omitting the last sections if time runs out. Evidence from homework and tests shows that there are enough hard ideas here for the students to benefit from hearing them in addition to reading them. Another method is to bring out the main ideas in discussion. Take the Salk vaccine field trial, for example. We present the background to the trial, as outlined in the text. Then we say:

> Suppose they gave the vaccine to everybody, and the incidence of polio went down. Would that show the vaccine was effective?

The class usually figures out why not. Then we present the design which puts the consent group in treatment and the no-consent group in control and ask about that. The first objection is almost always that the treatment and control groups are different sizes. After dealing with that, the class will figure out that the two groups will differ in some more important way, although they many not be able to say exactly how; we then explain that polio is a disease of hygiene (p4 of the text). Then we present the NFIP design (grade 2 in treatment, grades 1 and 3 in control), and ask for comments on that. Then we talk about running a proper controlled experiment, and ask the class whether the assignment should be done by the toss of a coin, or by expert judgment. Then we go on to talk about placebos and double-blinding; these ideas are hard to elicit.

If the class is too small or too large, discussion can collapse; however, we have had good discussions with classes ranging from 20 to 200 students. The length of the discussions has never been a problem; if time runs out we just drop some sections in the chapter, assigning them for reading. On the other hand, an instructor who wants to present additional material on design will find many examples in the exercises; others are cited in the footnotes.

PART II. DESCRIPTIVE STATISTICS

For students, descriptive statistics is much easier to understand than probability or inference, and it may be a more important topic. This part of the book is about descriptive statistics for one variable--the histogram, average, standard deviation--and their relation to the normal curve.

A first pass is made at the topic of measurement error, in chapter 6. This may seem out of place in an introductory course, but it embodies one of the great lessons of statistics: every empirical number is subject to error, whether it is generated in a physics lab, a market survey, or a census. If the number is determined again, it comes out a bit different. In fact, the variability in repeated measurements is a basis for judging the likely size of the error.

Chapter 3. The Histogram

The main object of this chapter is teaching students how to read a histogram, but we found this hard to do without also teaching them how to draw one. Now drawing a histogram--or any graph--is hard work. Students will need pencil, graph paper, and eraser; at first, they will have to be helped with the rudimentary mechanics, like laying out axes.

We originally tried to fudge the definition of a histogram, but kept getting caught in contradictions. Eventually, we were forced to follow the definition quite strictly, which is perhaps unusual in an elementary text. For us, percentages are represented by areas. In this setup, the height of a histogram shows crowding or *density*: it is percent per unit length. (The word "density" has a technical sound, and is downplayed for that reason.) The units of density, for instance % per $1000, are complicated; but we couldn't get around that. Our experience is that "% per $1000" goes down better than "%/$1000."

There are two advantages to the area approach:

- There is only one kind of histogram to deal with (other books move from "frequency" to "relative frequency" to density).
- The histogram can be matched up against the normal curve so that area under the curve becomes intelligible.

Histograms will be used a lot in this book, so it is important to get the students used to looking at them.

Chapter 3 also introduces the idea of a *variable*, with the following classification:

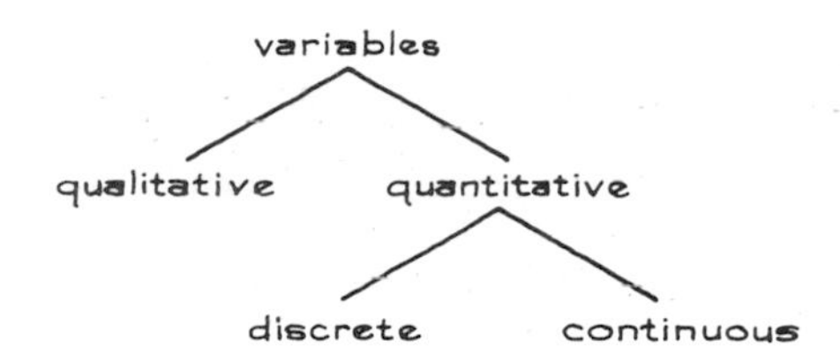

Our students didn't seem to like this much, but then they didn't seem to like any distinctions. (Perhaps they lack the experience needed to appreciate the usefulness of the distinctions, and don't want to be examined on things they don't quite grasp.)

Notes on graphics. Liberal use is made of smooth curves to indicate the shapes of histograms (as on p32), and some students will need reassurance about this. The point of sketching the histogram is usually to show some qualitative feature, such as the weight in the tails. For this, a smooth curve is just as good as the histogram, and is easier on the eye. In general, the art work has been kept fairly informal, in the hope that working diagrams will not look too forbidding.

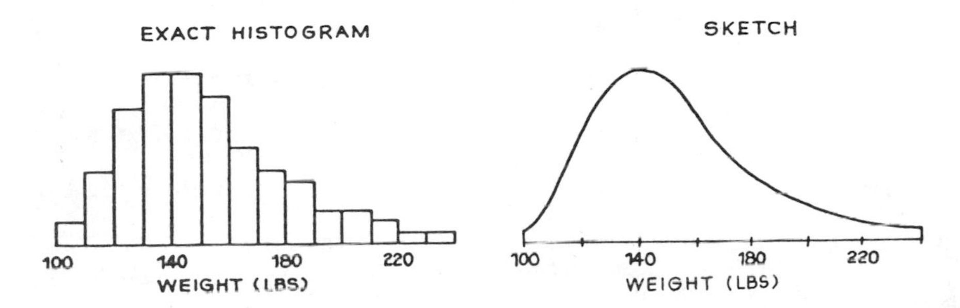

Notes on review exercises. Exercises 1–8 teach the interpretation of histograms: numbers 1 and 4 are crucial. Exercises 2 and 3 are for practice in drawing the graphs. Exercises 9–11 are hard: with number 9, students will prefer explanations in terms of any real factors--epidemics, immigration, whatever--to the statistical explanation in terms of digit preference.

Notes on data. Some instructors use data sets of their own to illustrate the statistical techniques discussed in the book; this works out well. Some do stem-and-leaf plots on small data sets before presenting histograms, and report good results from this approach.

Chapter 4. The Average and the Standard Deviation

The chapter focuses on interpreting these two statistics. "Standard deviation" is abbreviated to "SD," read "ess dee."

Variance is not introduced, for two reasons:

- Students get confused between Var and SD--"Is SD = $\sqrt{\text{Var}}$ or Var = $\sqrt{\text{SD}}$?"
- Var comes out in the wrong units, and the wrong order of magnitude.

For instance, American men average 170 pounds in weight, with an SD of 30 pounds. So the variance of weight is-- 900 square pounds. To a mathematician, taking the square root is an easy fix. However, we think it is quite hard to visualize the impact of a square root (or even a linear transformation), without actually doing the arithmetic. For instance, is 17 degrees Celsius warm or cold? If you live in a Fahrenheit world, you might reach for a calculator.

So we decided to focus on the SD, deferring the concept of variance to later courses. And even before presenting the calculation of the SD, the book explains the interpretation: the SD measures how far away, on the whole, the numbers are from their average. This interpretation can be fleshed out in the usual way:

- For many lists of numbers, about 68% of the entries are within one SD of average, and 95% are within two SDs.

The book points out that this rule isn't exact or universal. We hope it won't be misconstrued as slavish devotion to the normal curve: in fact, it works surprisingly well for many data sets that don't follow the normal curve at all (footnote 9 to the chapter).

The root-mean-square operation is presented in section 4, as a mathematical preliminary to computing the SD. In fact, taking the r.m.s. is a basic operation in statistics. For instance, it comes up again for the regression line (chapter 12). We used to introduce it there, but found that the students had a terrible time distinguishing between the r.m.s. error of the regression line and the SD of y. Moving the r.m.s. forward solved that problem. We have to admit, however, that it caused a new one: some students now confuse the r.m.s. and the SD. This is easier to sort out (exercises 9 and 10 on p69) but the instructor should be prepared to help.

Students may ask, "Instead of doing the r.m.s., why not just drop the signs and average?" We do not have such a good answer, except to say that the r.m.s. fits in better with the theory. Later in the course, instructors can explain that with large samples, it is the SD of the population which determines the asymptotic distribution of the sample average around the population average. Competing measures of spread, like the average absolute deviation from average, just won't do the job (footnote 9 to chapter 18).

The technical definition of the SD as the r.m.s. deviation from average is presented on p67. This reinforces the interpretation of the SD as a measure of the overall size of the deviations from average. Test results indicate that the prose teaches virtually all the students to calculate the SD correctly. But if not made to practice, they forget the algorithm within a few weeks.

The book only teaches one procedure for computing the SD. The other main one-- $\sqrt{\overline{x^2} - \overline{x}^2}$ -- is mentioned on p69. We used to teach this (as well as procedures for grouped data), but it only confused the students and made them learn less rather than more. They never seemed to believe that the two formulas would give the same answer, so they worried about which one to use, or combined them in unfortunate ways:

$$s = \sqrt{\frac{1}{n} \sum_{i=1}^{n} (x_i - \bar{x})^2 - \bar{x}^2}.$$

For us, alternative formulas represent a diversion from the main objective: teaching the students how to use the SD. After all, modern calculators make it less important for people to learn efficient computing algorithms--they just have to enter the data and push a button.

Notes on review exercises. Many of the exercises, like number 2 or number 13, focus on the qualitative ideas. Exercise 3, for instance, requires students to make a rough guess as to the answer: this forces them to think about the idea, instead of rushing to the formula and plugging in. Exercise 14 is hard, because students won't fit it into the cross-sectional vs. longitudinal framework. Exercise 10 is on the same point. The HANES data are cross-sectional, so the older people in the study were born earlier, are less well educated, and some of their skills have become obsolescent. Students often "explain" the results in terms of employment status: older people are less likely to be working. It turns out that controlling for this factor does not really change the shape of the curve of average income against age.

Notation. When working at the blackboard, we write "ave" and "SD." We do not use $\bar{x}$, s, μ or σ. When we did, students put so much energy into mastering the symbols that they had nothing left over for the ideas.

Which SD? The text defines the SD with n (the number of entries in the list) in the denominator, rather than n-1. The n-1 is introduced much later (section 26.6) as one of the modifications needed to handle small samples.

We felt that in the main line of exposition, there should be only one formula for the SD. To see why we went for n, consider the average of n draws made at random with replacement from the box [-1] [+1]. When n is reasonably large, this average will be in the range $-1/\sqrt{n}$ to $+1/\sqrt{n}$ with probability about 68%. We want this interval to be of the form $\pm\sigma/\sqrt{n}$, where σ is the SD of {-1,+1} . So, the SD of {-1,+1} has to be computed with 2 in the denominator, not 2 - 1 = 1. In other words, when calculating the SD of a population in order to determine the asymptotic behavior of the sample average, the right denominator is n.

The conventional argument for n-1 is that $\frac{1}{n-1}\sum_1^{n-1}(x_i-\bar{x})^2$ is unbiased. So it is, unless a regression is involved, in which case n-p is needed. And even without regressions, the minute someone takes square roots to get the SD, bias comes back. We know it looks old-fashioned, but n is the right denominator for present purposes.

Chapter 5. The Normal Approximation for Data

This key chapter ties together histograms, the average, the SD, and the normal curve. The passage on pp76—77, which justifies the 68%—95% rule, is difficult to teach. For instance, take figure 2. The shaded area under the histogram between 61 inches and 66 inches represents the percentage of women with heights in that range, which is the interval within 1 SD of the average. By inspection, the shaded area is about equal to the area under the normal curve between -1 and 1. This last area is 68%, justifying the rule. However, when asked, "What does the area under the histogram between 61 inches and 66 inches represent?" many students will respond "68%." Their anxiety to get to the numerical answer shortcuts the logic. Review exercise 1 of chapter 3 is designed to help prevent this; also see review exercise 10 in the present chapter.

Our method for teaching the normal approximation is graphical. On the blackboard, we draw diagrams just like the ones in examples 8 or 9 on pp81–82. Unless pushed, students seem to resist drawing these diagrams (or any others). Then later on in the course, with more complicated problems, they lose track of which areas they want. The diagrams help.

Section 4 takes up percentiles. It also shows that many histograms are far from the normal curve, a point which comes up again in section 6.3. The point is important, because some students take the word "normal" very literally indeed (p84 of the book). For this reason, we try to avoid phrases like "normal histograms," saying instead "histograms which follow the normal curve."

Section 5, on finding percentiles for the normal curve, will be tough going for some students. This material is used again--glancingly--in part III.

Note on terminology. In this book, a histogram "follows" the normal curve if it is close to the curve.

Notes on review exercises. Exercises 8–12 are hard. To help students work exercise 12, we ask them to mark (by eye) the average on the histogram, as well as the region within one SD of the average. Then we get them to work out ave±SD, using the values given in the problem.

Chapter 6. Measurement Error

Students may confuse chance error and bias. They may also need help in seeing that the SD of a series of repeated measurements gives the likely size of the chance error in each one (pp94–95 of the book, and chapter 24).[1]

The text has the equation

individual measurement = exact value + bias + chance error.

Some tact is needed when presenting this, because many students want to solve for the unknowns on the right, and feel cheated when they discover this to be impossible. The equation is a useful conceptual tool. Even though the unknowns cannot be precisely determined, they can often be estimated quite well.

Outliers are discussed in section 3, emphasizing the point that many histograms just do not follow the normal curve.

Notes on review exercises. This set covers most of the ideas in parts I and II. Exercises 11–12 prove difficult

[1] "Likely size" just means to an order of magnitude: chance errors comparable in size to the SD are common; chance errors several times larger than the SD are quite rare. A similar point comes up again on p36 and p40 of this manual.

for students who want to operate formally with the SD, instead of seeing it concretely as a measure of spread. Such students think the SD should stay the same. To help, we tell them to think about having all the men and women in a classroom, then sending the women out; what does this do to the spread in heights? Exercises 14–15, on design issues, are hard. In exercise 15, the tables are quoted from the source, and the numbers really do not add up. Exercise 16 illustrates digit preference; also see exercise 9 in chapter 3. Exercise 17, like many others in the book, may provoke students who want a self-contained mathematics course--free of background facts.

Chapter 7. Plotting Points and Lines

The presence of this chapter may be a bit of a shock. However, only half the elementary statistics students at Berkeley can plot a point on a graph. Some teachers may want to spend an hour on this chapter. Our approach is to review points and lines as we cover part III; students who need extra help can read chapter 7 by themselves.

PART III. CORRELATION AND REGRESSION

This part of the book is about bivariate data--scatter diagrams, the correlation coefficient, and the regression line. The treatment is purely descriptive. Many teachers may wish to postpone or even skip part III. It is possible to move directly from part II to chapter 13 (probability), and then to part V (chance variability). It is also possible to do just chapters 8 and 10 from this part of the book. However, part III does follow naturally from part II, and it is easier for the students than parts IV–VIII.

In chapter 8, the correlation coefficient is presented as a key descriptive statistic needed to summarize the relationship between two variables. Then r is used to get the regression line going in chapter 10, and to determine the spread around the line in chapter 11. We used to do the regression line before introducing the correlation coefficient, but this proved too mathematical for the audience, because it depended on the connection between a line and its equation--a famous sore point from high school.

Chapter 8. Correlation

The main job is to teach students how to read (and draw) scatter diagrams. Then, *association* is discussed carefully. (Students who work exercise set A on pp115–17 will get comfortable with these ideas.) Next, the correlation coefficient is interpreted graphically, as measuring clustering around a line. (It is clearer for the students to say that r measures clustering, rather than spread, because as r goes up to 1, clustering increases and spread decreases.) Scatter diagrams are summarized by the five statistics on p118; the warning about outliers or nonlinear association is deferred to section 9.3. An algorithm for computing the correlation coefficient is presented on pp124–25. (All the comments about computing the SD apply here too, in spades.)

Note on terminology. We found it very helpful to introduce two nonstandard entities:

- The *point of averages* (ave. of x, ave. of y) picks out the center of the scatter diagram (p117).
- The *SD line* indicates the drift of the scatter diagram (p123). It goes through the point of averages, and its slope is (SD of y)/(SD of x); the sign is the same as that of r. (If r is 0, either sign can be used.)

Many nonstatisticians (and some statisticians) who fit a line to a scatter diagram by eye will approximate the SD line rather than the regression line. The contrast between the two is the regression effect (section 10.4). For us, the main point of the SD line is to help explain regression.

Notes on review exercises. The graphical interpretation of r is covered by exercises 1 and 2; the computation of r, by exercise 9. Exercises 3, 5 and 12 are about association; exercise 5 is a bit tricky. Exercise 11 is about the SD line. The remaining exercises (4,6,7,8,10) test the interpretation of r as a measure of linear association. Exercise 8 is hard: to help students work it, we ask them to plot some data points.

Chapter 9. More about Correlation

Section 1 explains that r is a pure number, which is invariant under scaling, and symmetric in x and y. (The last point has some force later, because students will interpret r as a measure of causation.) Since r is invariant under change of scale, "clustering" must be interpreted relative to the SDs. This is somewhat delicate, as indicated by figure 3 on p137. Section 3 explains that r may not be useful if there is a strong nonlinear association, or outliers.

Section 4 discusses the fallacy of ecological correlations. (The term "ecological" is mysterious, and is downplayed in the text.) This may be a controversial section, because many investigators in the social sciences use ecological correlations without batting an eye.

For many students, a real intellectual effort is needed to compute r. They conclude that it must be a very powerful tool. It is; but there are limits, and section 5 points some of them out.

Notes on review exercises. Exercise 4 will be difficult for many students, but exercise set B on pp138–39 paves the way; also see exercise 11 on p99. Exercise 9 is about cross sectional vs. longitudinal studies. Exercises 11–13 are hard.

Chapter 10. Regression

Section 1 presents a verbal equivalent of the regression equation for estimating the average of y from x: If x goes up by one SD, on the average y does not go up by a whole SD, but only by part of an SD, namely, r × SD of y. Section 2 demonstrates how good the regression method is, using the graph which displays the average of y against x. In the text, this is called the *graph of averages*.

Section 3 takes up regression estimates for individuals, along with percentiles (which are a bit difficult). The material on percentiles can be skipped, although review exercises 6 and 7 cannot then be assigned.

The regression fallacy is discussed in section 4. This is the most interesting, and difficult, idea in parts II and III. When x goes up by one SD, most people want y to go up by a full SD too. The fact that it doesn't is the regression effect. The text explains that the regression effect is due to the spread of the scatter diagram around the SD line: figure 6 on p161. People resist this statistical explanation, and want some real cause for the regression effect: that is the regression fallacy.

The regression effect is implicit in section 1, but there it is kept in a very low key; we wanted the students to learn the mechanics before confronting the mystery.

Section 5 explains that there are two regression lines, one for y on x, one for x on y. There is ample room for confusion here. For example, in figure 9, the regression line of height on weight is steeper than the SD line; how come? (Answer: the dependent variable is plotted on the horizontal axis.)

Notes on review exercises. Exercises 3—5 and 7 demand a real understanding of the regression effect, and are difficult; exercise 9 deals with the two lines.

Note on the regression equation. The equation behind the prose treatment is

$$\frac{y - \bar{y}}{\text{SD } y} = r\,\frac{x - \bar{x}}{\text{SD } x}.$$

We used to teach the equation. Students would ask what r meant, as well as SD x and SD y, to say nothing of $\bar{x}$ and $\bar{y}$. This was fair enough. Then they would ask what x was, at which point we got a bit discouraged. Finally, they would ask what y was. We gave the equation up as a bad job.

Note on terminology. The "graph of averages" is not a standard term, but we found it useful in discussing the regression line. In principle, this graph depends on how finely you subdivide the x's.

Chapter 11. The R.M.S. Error for Regression

This chapter introduces residuals, as well as the formula for the r.m.s. error of the regression line. Students may want to know why they need both r and the r.m.s. error: one answer is that r is in relative terms (relative to the SDs) while the r.m.s. error is in the same units as y. Residual plots are taken up in section 3, although their power only becomes apparent with multiple regression.

The definition of "homoscedastic" on p180 is a problem for some students. As far as they can see, the scatter diagram in figure 9 (p179) shows more spread in a strip over 68 inches than in the strips over 64 or 72 inches. They are using range to measure spread; and the range is bigger in the middle of the diagram, because there are more people there. (This is taken up in the text when homoscedasticity is defined; also see exercises 2 and 11 in set D, chapter 4, as well as exercise 4 in set E.)

For "football-shaped scatter diagrams"--bivariate normal distributions--section 5 shows how to calculate the distribution of y when x is confined to a narrow strip: of course, that is the conditional distribution of y given x.

The calculation is a bit intricate. Students will have a hard time connecting the "global" r.m.s. error and the new SD, which is "local." Exercises 2 and 3 on p182 are designed to make the connection. (Many students will ignore the heteroscedasticity in exercise 3, and just do the arithmetic; also, exercise 4 on p182 is a shocker--the regression effect in acute form.)

The focus of chapter 11 is descriptive, not inferential. The r.m.s. error measures the spread of the points around the regression line. The chapter does not consider uncertainty in the position of the regression line, which increases with distance from the point of averages. Despite the relatively narrow focus, chapter 11 will take some time to teach.

Notes on review exercises. Exercise 9 is about measurement error; a common student response to (a) is "to see the regression effect." (Charitably interpreted, this isn't so bad; the point is to see the spread around the SD line.) Exercise 11 requires contact between the statistics course and life, which may disconcert some students.

Chapter 12. The Regression Line

The regression equation is presented in section 1, as an aid to computing (the exercises were set up with this in mind). The slope and intercept of the regression line are interpreted as descriptive statistics, with a warning about confounding. Section 2 discusses fitting a straight line to data in order to estimate the slope and intercept of an ideal linear relationship, and makes the point that the regression line minimizes the r.m.s. error. This material will not be easy. Section 3 restates the difficulties in drawing causal inferences from slopes.

Review exercises 9 and 10 are hard, because students do not recognize the regression line from its description. We encourage them to sketch a scatter diagram for the income-IQ data, find the point of averages, draw the line defined by the exercise, and mark the strip corresponding to children with the given IQ. Then we ask the students to find the center of that strip.

PART IV. PROBABILITY

As probabilists, we like the subject a lot; but students find it confusing. And whatever the advocates of the new math used to say, sets and functions make things worse for beginners. We also found that very little probability is needed to handle the statistics presented later in the book. So we went back to a more primitive approach. Chapter 13 handles the basics--independence--and sometimes we skip the rest of part IV; section 14.1 on counting and chapter 15 on the binomial distribution help just a little, when setting up probability histograms in chapter 18.

Chapter 13. What Are the Chances?

Sections 1 and 2 explain the frequency interpretation of chance. We could only afford one interpretation, and this seemed to be the smoothest. We hope that colleagues who belong to other schools of thought will not be too offended. Section 3 presents conditional probabilities; example 2a) responds to students who have trouble thinking about the chance that the second card dealt from a deck will be the queen of hearts: "What's the first card?" So we try to explain what an unconditional probability is, first.

Section 4 does the multiplication rule, and independence--the key idea for later--comes in section 5. *Collins* is discussed in section 6, showing that the assumption of independence matters. This opens one of the major themes of the book: When does the theory of chance apply? What happens if it is used in a situation where it does not apply?

Notes on review exercises. These exercises are simple and qualitative, in order to encourage the students to think about the issues. (Displays of professional cleverness in calculating things are especially disastrous when teaching probability; the students just wonder how they'll ever manage.) Some students will have trouble with exercise 3: we explain that it is harder to jump two hurdles than one.

Exercise 10 asks for the chance of not getting 4 sixes on 4 rolls of a die. Many students will answer this by calculating the chance of getting 4 non-sixes, $(5/6)^4$. To help such students sort things out, we ask them if the dice can land so as to get some sixes, but not 4 of them; we try to extract very specific answers, like:

⚅ ⚂ ⚀ ⚅

Exercise 11 prepares for expected values and chance models.

Notation. On the blackboard, we write fragments like

"chance of heads"

or

"ch. of ace on 1st roll and ace on 2nd roll."

We try to avoid "P(A)," "P(heads)," "A ∩ B," "red ∪ black."

Chapter 14. More about Chance

Thinking about the set of all possible ways that a chance experiment can turn out is a very useful technique, and section 1 presents it. Section 2 presents the addition rule. Example 7 is non-trivial, because many students want the chance of getting at least one ace in two rolls of a die to be 1/6 + 1/6. The double-counting argument is a bit abstract; at this point, the sample-space representation of chances would be quite powerful, and figure 1 is a reasonable substitute.

Section 3, on the paradox of the Chevalier de Méré, is an example of how to compute probabilities using the method of complements. Students find this a bit too clever: instead of being impressed that the problem can be done at all, they are annoyed at not having a simpler way to do it--and worried about what is to come on the final.

The focus in chapters 13 and 14 is qualitative, getting across the new concepts of independent and "mutually exclusive" events, and trying to separate them. Students have a hard time with these two ideas. After all, both express ideas of unrelatedness; there is a natural temptation to merge any two new ideas: and another temptation to think that if one doesn't apply, the other must. Exercises are designed to ward these temptations off, with partial success; and see the homily on pp228–29.

Of course, basic probability really does involve fractions, and this may demoralize some students.[1] The rest of the book features decimals, which are easier.

Notes on review exercises. Exercises 2 and 3 teach that two chances are better than one; after all, that is why students like midterms. Exercises 4 and 11 help the students distinguish between "independent" and "mutually exclusive" events. Exercises 12 and 13 prepare for expected values and chance models.

Chapter 15. The Binomial Coefficients

This chapter explains how to calculate binomial probabilities. We skipped the derivation of the coefficients; some instructors may wish to do this in class.

Notes on review exercises. Exercise 2 helps separate the binomial formula from the previous methods for computing

---

[1] According to the NAEP, only 68% of the seventeen-year-olds in school in the United States can add 1/2 and 1/3. Berkeley students can add fractions; even for them, however,

$$1/2 \text{ of } 1/3 = 1/2 \times 1/3 = 1/6$$

is rote learning--"of means times." For proof, see the pre-test results.

probabilities. Exercise 8 smuggles in the sign test. The sign test is an attractive introduction to significance-testing, but there is a hitch. In this context, students want to get the P-value by computing the probability of the observed outcome; they do not like tail probabilities at all, and who can blame them? We prefer to deal with this issue in a setting where the chance of any particular outcome is too small to be interesting (section 26.1).

PART V. CHANCE VARIABILITY

One famous difficulty in teaching elementary statistics is getting across the idea that the sample average is a random variable. Randomness, after all, is quite a complicated idea. It is easily overwhelmed, either by the definiteness of the data, or by the arithmetic needed to calculate the average.

In our experience, the most intelligible short explanation goes something like this:

> You took a sample and computed the average. That is a number. But it could have come out a bit differently. In fact, if you did the whole thing all over again, it would come out differently.

This variability is the key point to get across, and it tends to be obscured by the technical sound of the phrase "random variable." As a result, we have given that phrase up (and many other hallmarks of civilization too).

For the phrase, at least, there is a good substitute: drawing at random from a box of tickets, where each ticket has a number written on it. This may seem crude, but conveys a clear image.

To bring variability into sharper focus, we use the idea of *chance error*. For instance, when we talk about the sample average (chapter 23 in part VI), we tell the students:

> Draw some tickets at random from a box, and take the averages of the numbers you get. This will be close to the average of all the numbers in the box, but it will be a little bit off. This amount off is *chance error*:
>
> average of draws = average of box + chance error.

How big is the chance error likely to be? This question is answered by a number we call the *standard error* (abbreviated to SE, read "ess eee"). The upshot is that the average of the draws will be around the average of the box, give or take an SE or so.

Technically, a "chance error" is the difference between a random variable X and its expected value E(X). The "standard error" of X is $\sqrt{E\{[X-E(X)]^2\}}$. (At the risk of the obvious, the formula disappeared from the text at a very early stage.)

"Standard error," of course, is not the usual term; most authors use "standard deviation" both for data and for random variables. In our experience, however, students have a lot of trouble separating the standard error for the sample average from the SD of the sample. Calling the two by the same name makes it hopeless. So in this book we are quite rigid:

- The SD is for data.
- The SE is for random variables.

Drawing tickets from a box, chance variability, expected values, standard errors, the normal approximation....That is a lot of ideas. It takes times to get them across, and it is very hard to deal with them adequately in the middle of a complicated discussion on sampling.

So we develop these ideas first, in part V, focusing on the sum of draws made at random with replacement from a box.[1] We start with the sum because chance variability is easier to recognize for sums than averages.

We handle chance variability with more care than is common in elementary books. Our pedagogical motives should be clear by now: the ideas are hard, and need time to sink in. But we also have to admit an ideological motive. We think that statistical inferences should be based on explicit chance models. The argument is presented in the text; sections 21.4—5, 22.5, 23.4, 24.4, and chapter 29.

Now students are busy people, slightly cynical, with a definite short-term goal: passing the final. Their previous mathematical education stresses arithmetic procedure, not logical deduction. It is useless to tell them, "Statistical inferences should be based on chance models." This is empty rhetoric, with a lot of fancy words; and no sensible exam question can be based on it. We want students to take chance models seriously, so we spend course time on the topic. We also have exercises where getting the model wrong leads to the wrong answer--and losing points.

---

[1]Technically, this is a substitute for that famous sum of independent, identically distributed random variables. We are sacrificing some generality: our random variables only take finitely many values, with rational probabilities. That's enough.

A final remark. Part V is independent of part IV. Instructors who want to spend the minimum amount of time on "pure probability" should, in our opinion, skip part IV but do part V. Part V only takes three or four hours of class time, and it is a very good investment.

Chapter 16. The Law of Averages

Students often think that with a good sample, the sample percentage will equal the population percentage. This makes it difficult for them to appreciate the standard error calculations in part VI. Part of the trouble is that they don't understand chance variability. So section 1 of chapter 16 takes this up. We have a coin. On each toss, it is as likely to land heads as tails. Now we toss it 10,000 times. Are we likely to get exactly 5,000 heads? Surely not. As the number of tosses goes up, the difference between the number of heads and the expected number tends to get larger and larger in absolute terms, that is, as a number. However, the difference tends to get smaller and smaller in percentage terms, relative to the number of tosses. For many students, this distinction is new and difficult. It is central to the careful discussion of the law of averages in section 1. This section also discusses the concept of chance error, with the equation

number of heads = half the number of tosses + chance error.

The *likely size* of the chance error is used informally in the text. (The technical equivalent is the standard error.)

The balance of the chapter is spent setting up box models and introducing the sum of the draws from the box. A box model consists of draws made at random from a box of tickets; each ticket in the box shows a number. The chance variability in coins, dice, roulette wheels (and later, sampling processes) is related to the chance variability in draws from a box. Eventually, this produces real economy of thought: there is a general theory, instead of a lot of special cases. At first, students find this approach quite strange, but they get used to it very quickly.

Many examples in this chapter are based on gambling at roulette: the sum of the draws from the box corresponds to the net gain. For instance, take example 1 on p257. The net gain in 100 plays at roulette, staking $1 on a single number at each play, is like the sum of 100 draws from the box:

1 ticket	$35	37 tickets	-$1

The phrase "is like" has a precise technical meaning: the net gain and the sum have the same probability distribution. Of course, we do not insist on this in the text, but make the point through problems like exercise 6 on p255 or 3 on p259.

Students find the gambling interesting, although a bit technical. (One touchy point is adding up negative numbers.) It is a digression from the mainline statistical issues. However, setting up a proper model for a mainline statistics problem is hard. Setting up a model for roulette is much easier, and it's good practice. As we tell the students, the first step is to write the box down. (Of course, you can quickly generate a lot of free-floating boxes; nobody said this was an easy subject to teach.)

Notes on review exercises. Exercises 1 and 2 test the distinction between absolute and relative errors, and are hard. Exercise 2 is easier for the students when translated into a problem about coin-tossing.

Chapter 17. The Expected Value and Standard Error

This chapter presents the formulas for the expected value and standard error for the sum of draws made at random with replacement from a box. The first idea is that the sum of the draws from a box will be around its expected value, but will be off by a chance error:

sum = expected value + chance error.

The likely size of the chance error is given by the SE for the sum. As we write over and over again on the blackboard,

The sum of the draws will be around _____
give or take _____ or so.

(The downside: some students will later view expected values as random variables, computed up to some margin of error; exercises 5–7 on p388 try to ward this off.)

We tell the students that chance errors of an SE or so in size are fairly common, but chance errors bigger than several SEs in size are very unusual.

The SE for the sum of draws made at random with replacement from a box is computed by the square root law (p265) as

$$\sqrt{\text{number of draws}} \times \text{SD of box}.$$

Students need help seeing what the square root means: when the number of draws goes up by a factor of 100, say, the SE for the sum of the draws only goes up by the factor $\sqrt{100} = 10$. In particular, as the number of draws goes up, the SE for the sum goes up in absolute terms, but goes down relative to the number of draws.

When the number of draws is large, the normal approximation can be used (section 3), although a full discussion is postponed to chapter 18.

As a matter of style, it is wise (though cumbersome) to write "SE for sum," not just "SE." We try to make the students do this too. Later on, we will have both the SE for sums and the SE for averages. Students will want to merge these two entities. Insisting on full names helps prevent this.

Many boxes in gambling problems (roulette, for instance) have only two kinds of tickets, and there is a short cut formula for the SD of the box. More technically, if $P\{X=a\} = p$ and $P\{X=b\} = 1-p$, the SE is $|a-b|\sqrt{p(1-p)}$. In words, this formula appears on p271.

We attempt to treat standard errors in a unified way, tracing everything back to sums. In section 5, a coin lands heads with probability p and is tossed n times: what is the standard error for the number of heads? This problem fits into the general framework of sums by the 0-1 coding trick, counting heads as 1 and tails as 0. The number of heads is like the sum of n draws made at random with replacement from a box where the fraction of tickets marked 1 is p, and the fraction of 0's is 1-p. The SE for the sum is, of course, $\sqrt{n}$ × the SD of the box: now use the short cut.

Unfortunately, the 0-1 coding isn't so simple, in part because adding up 0's and 1's only seems sensible to mathematicians. So the section goes through the coding in some detail. The students do have trouble remembering to put 0's and 1's on the tickets. This isn't so bad with coin-tossing: some numbers are needed, 0 and 1 seem reasonable. It is harder when rolling die and counting the number of 6's, still harder when taking a sample and counting the number of people with incomes over a given level. In these examples, the students are thinking about some vivid quantitative variable already: 0's and 1's look pale by comparison. The "classifying and counting" slogan is to help in overcoming this difficulty, and so is the cartoon on p275.

Section 6 closes by relating the law of averages to the square root law. It is the square root which makes the SE for the number of heads go up in absolute terms, but down in relative terms. Chapter 17 has a lot of material, and it may spill over into a second lecture. (On the other hand, chapters 16 and 18 go fairly quickly.)

The downside:

(i) When computing the SD of a 0-1 box, students insist on the factor "1-0" in the formula

$$(1-0) \times \sqrt{p(1-p)}.$$

They love substitution; it's what they've been trained to do in math courses.

(ii) Students may automatically change to 0's and 1's, even for quantitative data. (The crunch comes in part VIII.) To help students use 0-1 boxes only when needed, review exercise 5 in chapter 20 is on quantitative variables, as is number 8 in chapter 21, even though chapters 20 and 21 are about qualitative variables. Conversely, review exercise 5 in chapter 23 involves qualitative data.

Notes on review exercises. Some students have trouble getting started on exercise 4: the connection between percentages and probabilities may be problematic. Number 7c) sometimes provokes this aggravating reply:

> With a large number of tosses, the chance error should be large in absolute terms but small in percentage terms, so option (i) is better.

Exercise 9 will be difficult for students who think that two games with the same expected value must offer the same chance of winning; this exercise should demonstrate why the SE is needed. (For a preview, see exercise 5 on p273.)

Chapter 18. The Normal Approximation for Probability Histograms

We introduce probability calculations for sums through the normal curve: when the number of draws is large, there is about a 68% chance for the sum to be within one SE of the expected value, and so on. This topic is broached in chapter 17, and discussed fully in chapter 18.

The first idea in chapter 18 is a probability histogram: a graph which represents chance by area. These histograms are drawn *deus ex machina*--by the computer. However, we find them an order of magnitude easier to use in the classroom than a notional list of all possible samples. Probability histograms are introduced in figures 1 and 2 as the limit of empirical histograms (from simulations). The reason for thinking about products, as in figure 2, is to see that not everything is normally distributed: the normal curve is tied to sums. Students should work exercise set A, to pin down the interpretation of probability histograms.

Sections 3—4—5 present a "local" version of the central limit theorem: The probability histogram for the sum of a large number of draws from a box will follow the normal curve very closely. However, as the chapter points out, if the distribution of tickets in the box is highly skewed, then many draws may be needed before the approximation takes hold. (This will cause some test anxiety--how can they tell when it is safe to use the curve?)

A new technique is introduced in section 4, to estimate the chance that the sum will take on some given value--the "continuity correction." The technique is easy to teach, but the name is confusing, so we dropped the name. Similarly, we avoid using the term "central limit theorem." Theorems are what they never understood, and there is a built-in conflict between "central" and "limit."

Some instructors are troubled by the approach in chapter 18, because they want to get to the "global" central limit theorem: a sum will be in an interval with probability close to the corresponding area under the normal curve. In our experience, students see immediately that if the probability histogram for the sum is close to the curve, areas under the histogram--probabilities--must be close to areas under the curve. This is discussed in section 4. The local theorem does imply the global one, both intuitively and formally. And in fact, the students do learn how to estimate probabilities using the normal curve, as in review exercise 2.

With our approach, probability histograms have to be put into standard units before matching them to the normal curve. This scaling is done in figure 3 on p289, which is like figure 2 in chapter 5. The elided difficulty is non-trivial: Is the density of a+bX equal to f(a+bx)? or is it f((x-a)/b)? Scaling is our substitute for the equation

$$P\{S_n < x\} = P\left\{\frac{S_n - n\mu}{\sigma\sqrt{n}} < \frac{x - n\mu}{\sigma\sqrt{n}}\right\}$$

In our experience, this equation is a loser.

Note on the SD. As footnote 9 to the chapter explains, the normal approximation shows why the SD is so useful. The shape of the probability histogram for the sum of a large number of draws from a box depends only on the average and SD of the numbers in the box. Other measures of spread, like average absolute deviation from average, have very little to do with it.

Notes on the review exercises. Exercise 3 tries to reinforce the idea that the histogram gives the exact answer, and the normal curve is just an approximation. Since the probability histogram is a difficult idea, students will confuse it with the histogram for the data--the draws from the box: Exercise 5 may help the students with this distinction. Exercise 6 previews hypothesis testing. Exercise 12 reviews the graphical interpretation of the central limit theorem, in terms of area.

PART VI. SAMPLING

Chapter 19. Sample Surveys

There are a lot of ideas about sampling which are obvious to statisticians but not to others, and are well worth teaching in an elementary course. For example:

- Some samples are terrible.
- The method used to draw the sample matters.
- Handpicking the sample to get a representative cross-section tends not to work very well.
- Haphazard selection may be even worse.
- The best methods for drawing a sample involve the planned introduction of chance.
- If the non-response rate is high, the results can't be trusted.

Jumping straight into the calculations prevents the students from coming to grips with the basic ideas. That is why chapter 19 opens with a qualitative discussion, pinned to historical examples like the *Literary Digest* poll's choice of

Landon (section 2), and the Gallup poll's "election" of Dewey (section 3). Probability methods are discussed in section 4, and their success is documented in section 5.

Elementary books (ours is no exception) concentrate on simple random sampling. Of course, the technical meaning of "random" is quite a bit more specialized than the usual meaning:

> "Without definite aim, direction, rule, or method."
> --*Webster's*

An effort is required to make students appreciate the technical meaning of "random." We take our best shot in sections 19.4 and 20.1; review exercise 5 on p325 may reinforce the point.

Once they know what the terms mean, students think that with a simple random sample, the sample percentage is very likely to equal the population percentage. (They are capable of thinking so, yet going on to compute 95% confidence intervals in response to word problems.)

Chapter 16 was designed to prevent this confusion, and section 19.8 continues the work. Again, the chance-error language creates the image of the sample percentage coming close to the population percentage, but missing by a little:

sample percentage = population percentage + chance error.

(There is no bias term with simple random sampling.)

Real sample surveys, of course, use methods much more complicated than simple random sampling. Our book faces up to this issue. *Multistage cluster sampling* is introduced in section 4; it will be discussed again in chapter 22. Section 6 points to some of the difficulties faced by the Gallup poll; and section 7 discusses telephone surveys.

Some of our teaching assistants confuse quota sampling with stratified sampling, and then wonder why we are attacking stratification. We aren't; and the two methods are very different, although they start out the same way. The crucial difference is that with quota sampling, the interviewer is free to choose respondents to make up the quota. For a stratified sample, the choice of sampling units within each stratum is done objectively, using chance.

Review exercises 7 and 8 are designed to ease the students into confidence intervals; but the connection may need to be pointed out.

Students seem to like chapter 19 very well, and they have little trouble with the exercises. They do have trouble with the terminology: *sample percentage*, *population*, *population percentage*, and *parameter* are all a bit remote.

Chapter 20. Chance Errors in Sampling

Section 20.1 reviews the definition of simple random sampling, and drives home the idea that the sample percentage will differ from the population percentage. Section 20.2 presents our version of $\sqrt{pq/n}$, except that the formula doesn't appear. This may seem a bit idiosyncratic, and we would like to explain why we moved from the conventional formula to our version.

The students seemed to find $\sqrt{pq}$ rather hard to swallow. So we taught them to make a model with 0's and 1's in the box. Since we were working in percents, the formula became

$$\text{SD of 0-1 box}/\sqrt{n} \times 100\%.$$

We presented it this way for several years, but there was still a hitch. The students were willing to compute an SE as SD × $\sqrt{n}$ in part V. When they hit part VI, there was a tremendous shifting of gears needed to compute the SE as SD/$\sqrt{n}$. Once they changed over, they stopped being able to compute the SE for a sum as SD × $\sqrt{n}$: they insisted on dividing. We tried hard to explain that there was one formula to use with sums and another for averages, but they wouldn't buy it.

Eventually, we decided to have only one formula: the SE for a sum. Everything else is worked out from that. For instance, section 2 gives an example where 400 people are chosen at random from a population consisting of 3091 men and 3581 women; the problem is to compute the SE for the percentage of men in the sample. When presenting this problem in a lecture, we begin by writing on the blackboard:

Percent of men in sample will be around ___
give or take ____ or so.

Then we proceed as follows:

Step 1. Set up a box. First we write an empty box on the board:

We ask how many tickets there should be in the box. (Many students will answer 400.) Eventually, we arrive at

3581 [0]'s 3091 [1]'s

The number of men in the sample is like the sum of 400 draws from this box.

The last is a key sentence: it connects the box to the problem. (If students can be persuaded to write this sort of sentence on homework or tests, they will be in relatively good shape.)

Step 2. Now the calculation can be made:

$$\text{expected value for sum of 400 draws} = 400 \times \text{average of box}$$
$$= 400 \times 0.46 = 184$$

$$\text{SE for sum of 400 draws} = \sqrt{400} \times \text{SD of box}$$
$$= \sqrt{400} \times \sqrt{.46 \times .54} \approx 10.$$

We pause to interpret the results: the number of men in the sample will be around 184, give or take 10 or so.

Step 3. Now convert to percents: 184 out of 400 is 46%, and 10 out of 400 is 2.5%. So the percentage of men in the sample will be around 46%, give or take 2.5% or so.

This works reasonably well for many students, but many others will just compute "the SE" using a formula--and have about one chance in four of picking the right formula out of the tool box. (Section 23.3 would help--if they read it.)

So far, we have been a bit sloppy about whether the draws are to be made with or without replacement. (Of course, when the sample is only a small part of the population, it makes little difference.) Section 3 takes this point up in some detail, and eventually comes up with the correction factor

$$\sqrt{\frac{\text{number of tickets in the box - number of draws}}{\text{number of tickets in the box - one}}}$$

In our opinion, this formula is a bit technical for elementary students, and pushing it too hard obscures the really interesting point: when estimating percentages, accuracy depends mainly on the absolute size of the sample, rather than size relative to the population. The section is organized to highlight the point, but the students may be quite hard to convince.

Notes on the review exercises. Exercise 4 covers the procedure for calculating the SE for a percentage, connecting it to the SE for a sum. Exercises 5, 6, 8, 9 require computing the SE for a sum; they help prevent the students from forgetting about part V. Exercise 7 tests the point that accuracy depends mainly on the absolute size of the sample, rather than the relative size. Exercise 10 is absolutely straightforward for students who write down the 0-1 box model; lazier students, however, will just use 0.9 for the SD and get the wrong answer. Setting up a model is a key step in making any statistical inference, but is a hard skill to teach. Exercise 11 previews hypothesis testing.

Chapter 21. The Accuracy of Percentages

This chapter contains the first technical treatment of inference from the sample to the population. Section 1 states the question to be answered: how accurate is an estimated percentage likely to be? (Before that, however, the section reminds the student of the basic problem--the estimate is apt to be a bit off.) Accuracy is determined by the SE. The estimate is likely to be about right, but off by an SE or so.

With apologies to Brad Efron, the procedure for estimating the standard error from the sample is "the bootstrap method:" substitution of estimates for parameters in the formula.[1] (In this chapter, the samples are large.) Many students have trouble with this, because they do not distinguish between what is known and what is unknown.

And the point is delicate. After all, there is a substantial shift from the last chapter to this one. For instance, suppose there is a town with 10,000 residents of voting age and unknown political preferences. To estimate the percentage of Democrats in the town, a simple random sample of size 100 will be used. Consider two strategies:

- Determine the political leanings of every one of these 10,000 people, draw 100 at random and take the percentage of Democrats in the sample.
- Draw 100 at random, determine their political leanings and take the percentage of Democrats in the sample.

The first is zany, the second very practical. But the usual standard-error calculation is made by thinking about the first process; the result is carried over to the second. Mathematically, this is fine. (The probability distribution for the sample percentage of Democrats is the

---

[1] In our context, this happens to be a special case of the bootstrap.

same in both setups.) Students may feel the jolt; we smudged our way past this in section 21.1.

The main worked example in the section (p346) is repeated here for reference, as we discuss the problem in teaching it.

Example 1. In fall 1987, a city university had 25,000 registered students. A survey was made that term to estimate the percentage who were living at home. A simple random sample of 400 students was drawn, and it turned out that 317 of them were living at home. Estimate the percentage of students at the university that term who were living at home, and attach a standard error to the estimate.

In working such examples, teaching assistants often demonstrate a natural desire for mathematical efficiency:

$$\frac{\sqrt{400} \times \sqrt{0.79 \times 0.21}}{400} \times 100\% \approx 2\%.$$

We resist, because the parts lose their meaning for the students:

- $\sqrt{0.79 \times 0.21} \approx 0.41$ is the SD of the box, estimated by the bootstrap procedure.

- $\sqrt{400} \times 0.41 \approx 8$ is the SE for the number of students living at home. There were 317 such students in the sample, and the 8 measures the likely size of the chance error in the 317. (See exercise 1 on p347.)

This chapter makes the transition from probability calculations to statistical inference, and here is one consequence. Students will not take us seriously if we tell them, in working the example, "the sample number will be around its expected value give or take an SE or so." After all, the sample number is right there in front of them--it _is_ 317. But the 317 is a little shaky, being based on a sample; the 8 tells us how shaky: and that is how we interpret the SE.

After dealing with standard errors, the chapter explains how to get confidence intervals for the population percentage at the 68%, 95%, and 99.7% levels by going 1, 2, or 3 SEs either way from the sample percentage. (The distinction between 1.96 SEs and 2 SEs, for instance, just didn't seem worth pursuing.)

The conventional frequency interpretation for confidence intervals is given in section 3; Bayesian colleagues are asked to temper justice with mercy. Even for a hard-bitten frequentist, this is a difficult passage to teach, because many students will want to say,

> There is a 95% chance that the percentage of Democrats in the town is between....

This is such a natural human hope that we couldn't bear to deal with it too harshly. The section does try to explain that the chance variability is in the sampling process not in the parameter.

Unfortunately, students find confidence intervals quite hard; in struggling with the complications, they are likely to lose track of the main point. So the section restates it, at the top of p353: the SE tells you the likely size of the amount off.

From our perspective, there is nothing wrong with omitting confidence intervals, and focusing on the SE as a measure of reliability; just be careful about homework assignments.

As noted before, the Gallup poll uses a complex multistage cluster sample, and $\sqrt{pq/n}$ just does not apply. This is hard: students want to analyze the data, which is right there in front of them; they do not want to pay attention to the process generating the data, which is much more remote. The point is tackled in section 4. Many elementary statistics books do not face up to the issue, and perhaps that is one reason why investigators run around computing $\sqrt{pq/n}$ in situations where the results make very little sense.

Notes on terminology. (i) We could not write the chapter without using the sample percentage-population percentage terminology, which is confusing to some students. The percentage of Democrats in the sample and the percentage of Democrats in the town are much more tangible, and the students pick up the idea through the examples.

(ii) We try to distinguish between the "true" standard error computed from the box, and the standard error estimated from the sample. (See exercise 4 on p384.) The latter is a "standard error of estimate," but this terminological elaboration would be too confusing.

Notes on review exercises. Parts a–b) of exercise 1 cover the basics; parts c–d) are supposed to remind the students that confidence intervals depend on the normal approximation (and see 5–6 in exercise set B on p350). Review exercises 5–6 are meant to teach the students not to use the standard error formula where it does not apply; exercise 4 stops them from solving such problems by a purely syntactic analysis. Exercise 8 is designed to review techniques from part V, and keep quantitative variables alive: 0's and 1's are not always the right move.

Chapter 22. Measuring Employment and Unemployment

Government estimates for the unemployment rate are prepared from the monthly Current Population Survey. This sample survey is discussed from the ground up. Such detail is unusual in an elementary text, but it consolidates the understanding of the material presented in the previous chapters, and gives the students a flying start on understanding any other large-scale survey. We do not hold the students responsible for details of the design. We mainly want them to learn that real surveys do not use simple random samples, so $\sqrt{pq/n}$ does not apply: the standard errors have to be estimated differently. The half-sample method is sketched in section 5. One conclusion is that the calculation for the standard error has to depend on the sample design; if this is unknown, or not well defined, sensible calculations are hard to make.

Many professionals are surprised to find that the complex design used by the Current Population Survey gives somewhat less accuracy (for some parameters) than a simple random sample. Although the stratification and the ratio estimation reduce sampling error, the clustering increases it; that is the explanation (p368). Of course, without the clustering nobody could afford to carry the Survey out. The real surprise, to us, is that the Current Population Survey is almost as accurate as a simple random sample; in many complex designs, the effective sample size is reduced by 15% to 50%. The Current Population Survey design is amazingly good.

Other statisticians are surprised that the ratio estimates used by the Survey are practically unbiased. One explanation is that the sample is very large, so the percentages have rather small SEs, and the ratio estimates are almost linear in the data.

Notes on review exercises. Part a) of exercise 1 tests understanding of ratio estimates (section 4) and part b) does labor force definitions (section 3); exercises 2–3–4 are about the half-sample method (section 5). Exercises 5 and 10 review definitions from chapter 19; exercise 6 makes the point that the SE depends on the sampling method. Exercise 7 is about confidence intervals. Exercise 8 is intended to review the basic interpretation of the SE. Exercise 9 is about interviewer bias (chapter 19). Exercise 11 makes the students focus on the difference between sampling with or without replacement. The hardest exercise in this section is number 12, which reviews probability histograms. There are two stumbling blocks: (i) seeing that the number of heads when 100 coins are tossed is like the number of heads when one coin is tossed 100 times; (ii) separating repetitions of tossing the coin within the group of 100 from repetitions of tossing the whole group. The figure on the next page may help.

Review exercise 12 in chapter 22. A group of 100 coins are tossed over and over again. The top panel shows data on the number of heads with 100 repetitions (i.e., 100 × 100 = 10,000 individual tosses). The second panel is for 1000 repetitions (1000 × 100 = 100,000 tosses); the third, for 10,000 (10,000 × 100 = 1,000,000 tosses). The bottom panel is the probability histogram.

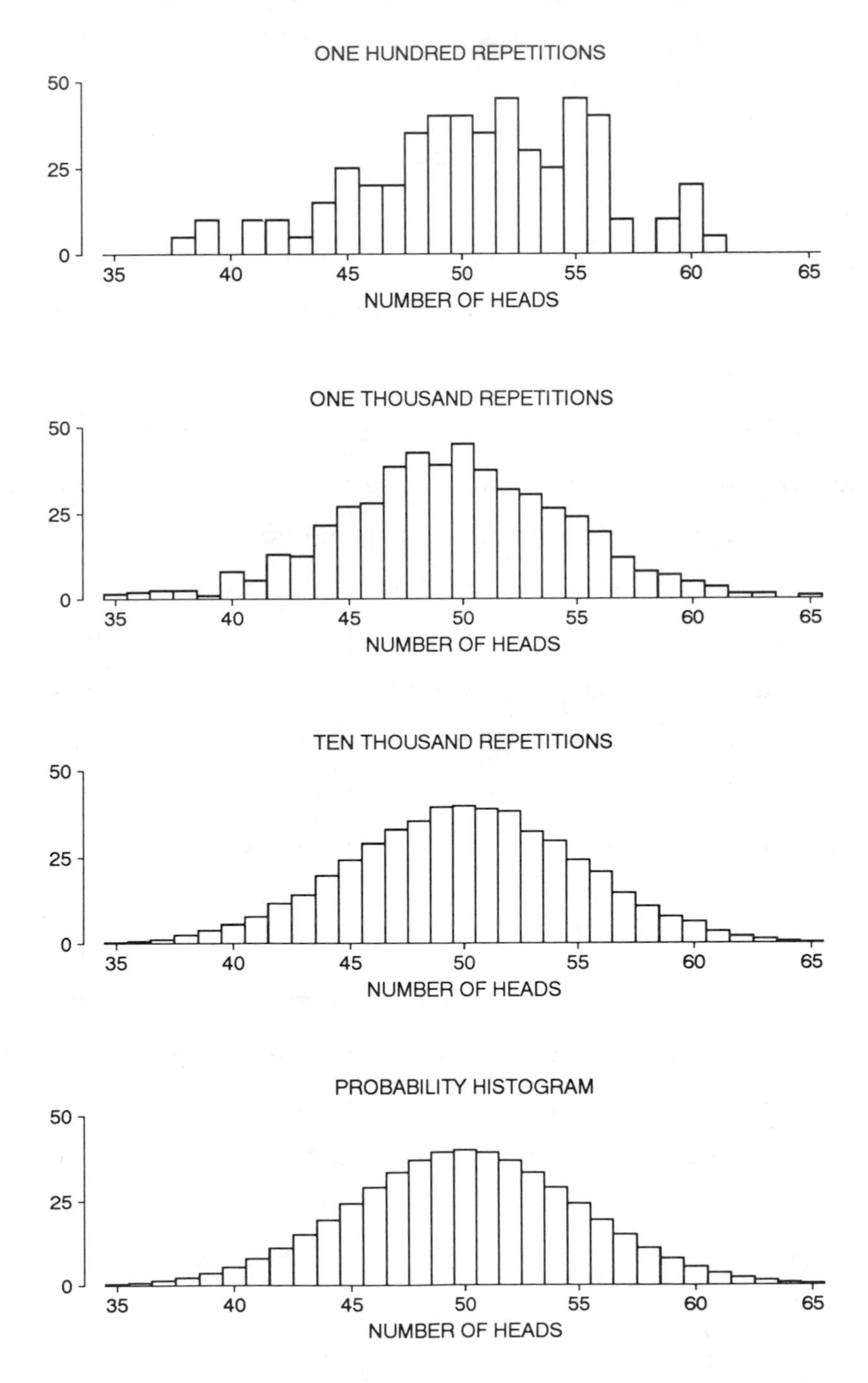

Chapter 23. The Accuracy of Averages

Section 1 explains how to calculate the standard error for the average of draws made at random with replacement from a box, by working back to the sum (p376 of the text). The interpretation is that the average of the draws will be around the average of the box, give or take an SE or so. Students handle this reasonably well, although by force of habit a few will go

SE for average of draws = (SE for sum/number of draws) × 100%.

Others will want to use the SE for the sum, with little sense that the order of magnitude is wrong.

The formula "$\sigma/\sqrt{n}$" appears only in the technical note on pp379–80 of the text; we do not teach it for reasons given earlier.

The application to inference is in section 2: with a simple random sample, the SE of the average is estimated by substituting the SD of the sample for the unknown SD of the box. Then, confidence intervals are obtained by going the right number of SEs either way from the average of the sample. (In this chapter, the samples are large: small samples are dealt with, by Student's t, in chapter 26.)

At this point, to mix a metaphor, a lot of very tough chickens will come home to roost. Many students are going (somehow) to want 0-1 boxes in section 2. Others will want to use the SE for the sum or the percent, rather than the SE for the average. Survivors will mix up the probability histogram for the average of the sample with a histogram for the data. Another confusion is between the SD of the sample and the SE of the average, so confidence intervals get interpreted as follows:

> 95% of the population is within 2 SEs of the average of the sample.

Other students fly over chapter 18, because they see no new techniques presented there; but in chapter 23, they have to come to grips with the central limit theorem. After all, how does the normal curve fit into a problem on educational levels, if the data are so far from normal? Our answer is figure 2 on p383. The normal curve is a good approximation not to the data, but to the probability histogram for the average of the draws.

The ideas in section 2 have all been introduced before, but they are difficult, and they interact in funny ways. We do not know how to slow things down enough, but the section tries to sort the issues out. Many students profit from studying figures 1 and 2; others get things under control by working exercise set B. Section 3, and exercise set C, will also help. However, the instructor should keep a close watch on the situation.

As noted before, the standard-error calculations presuppose simple random sampling, and the students are reminded of this in section 3. The calculations for confidence levels depend on the normal approximation; exercises 2–3 (p388) make the point.

Note on terminology. Students seem to find "sample average" a bit confusing: is it a sample of averages, or what? "The average of the sample" is better, and "the average of the draws" better yet. "Population" also tends to throw things off course: we find ourselves talking about "the box" and being understood better. For similar reasons, we don't use "parameter" very much.

Notes on review exercises. These exercises force the students to distinguish between the SE and the SD. They also make the students separate out the histogram for the sample and the probability histogram for the average of the sample. They teach that the calculations depend on the normal approximation, and on simple random sampling. So they are tough, but provide good diagnostics.

PART VII. CHANCE MODELS

In part VII, box models are used to study two topics: measurement error (chapter 24) and genetics (chapter 25). These topics are a bit unusual for an elementary statistics course; instructors who wish to skip them will find that part VIII was written with this possibility in mind.

Part VII is designed to reinforce the lesson that to make a good statistical inference, the investigator has to get the box model right.[1]

[1] Box models look special, because the draws (when made with replacement) are independent. However, they can be modified to handle dependence. Just for one example, a pair of dependent random variables can be modeled by drawing at random from a box of tickets, where each ticket shows a pair of numbers. See chapter 27.

Chapter 24. A Model for Measurement Error

With a large number of measurements of the same quantity, the standard error for the average of the measurements is estimated as in chapter 23. You start by finding the SE for the sum of the measurements--

$$\sqrt{\text{number of measurements}} \times \text{SD}.$$

Then, you divide by the number of measurements, to get from the sum to the average. As in the sampling context, there is room for confusion between the SE and the SD. The cartoon, and the discussion on pp396–97, try to separate these two quantities.

Despite the familiarity of the arithmetic, there is an issue in this chapter, and it is dealt with in sections 2-3. The procedure for computing the standard error is based on the square root law; the justification depends on viewing the measurements as the observed values of a sequence of independent, identically distributed random variables.

In our experience, that formulation would not convey much to students. We state the condition this way: the data are like the results of drawing (at random, with replacement) from a box of numbered tickets. In particular, if there is any trend or pattern in the data, the model does not apply (pp399–403). Of course, dependence between the measurements also rules the model out. Students can use this principle as a heuristic, relying on the ordinary meaning of "dependence."

(In many cases, the model fits measurement data rather badly. The investigator develops some notion of what the next measurement "ought" to be, based on the previous data, and tends to report this notion instead of the real measurement, destroying the independence. That kind of observer bias is eliminated by the weighing design used at the National Bureau of Standards. See footnote 6 to the chapter.)

Usually, one objective of measurement error models is to make a clean separation between the parameter being estimated (the "exact value" of the thing being measured) and the chance errors. There is a practical reason for this separation. For example, if repeated measurements are made by a certain process on a check weight, the variability in the results can be used to judge the likely size of the chance error in a measurement on another weight (example 5 on p405).

We set the model up with this in mind. There is a box of tickets, called the *error box*; each ticket in the box represents a possible chance error, and the average of the numbers in the box is assumed to be 0. Then, each measurement equals the exact value of the thing being measured, plus a draw with replacement from the box. This is the Gauss model for measurement error. (The name should not be taken to imply that the errors follow the normal curve.) In our somewhat primitive notation, this model can be put as follows:

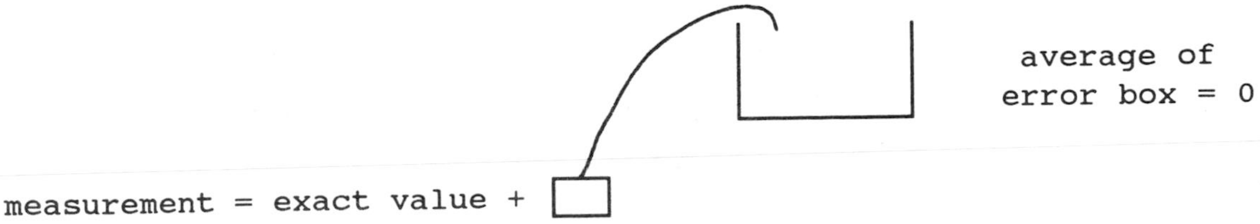

average of
error box = 0

measurement = exact value + □

More conventionally, the model would be stated as follows:

$$X_i = \mu + \varepsilon_i,$$

where the ε_i are independent, identically distributed, and have expectation 0.

The model is explained in section 3, and the procedure for calculating the SE is derived from the model. Bias--often a major problem--is taken up at the end of the section. (Up to this point, bias has been assumed to be negligible.) The role of the model in making inferences is summarized in section 4.

Notes on review exercises. Part a) of exercise 1 is the basic blurt; parts b–f) ward off various misinterpretations of confidence intervals. Exercise 5 is about the role of the model. Exercises 7 and 8 bring the SE for the sum back into play; of course, for the students, the first issue is to see that sums are involved. Exercise 10 might make them think about the normal approximation.

Chapter 25. Chance Models in Genetics

This chapter gives a brief account of Mendel's genetic theory, based on his experiments with peas. For statisticians, there is an interesting twist to the story: Fisher argued that Mendel's data were massaged to make the frequencies closer to their expected values (section 2). Fisher also showed that Galton's law of regression could be explained by Mendelian theory. One version of the argument is presented in section 3, but it is out of reach for most students.

The physical source of the randomness in Mendelian genetics is described in section 4. This is a tough story, but perhaps worth telling: one of the great strengths of the model is the precise description of the physical sources of randomness.

As we say in the text, this chapter is included for two reasons:

- Mendel's theory of genetics is great science.
- The theory shows the power of simple chance models in action.

PART VIII. TESTS OF SIGNIFICANCE

Chapter 26. Tests of Significance

Testing is an important topic, but hard to teach. The basic idea of the z-test is easy: if an observed value is too many SEs away from its expected value, something is wrong. But students find the vocabulary bewildering, and the implicit double negative is hard to follow: the investigator proceeds by rejecting the opposite of what he wants to prove. Our objective was to teach some of the conventional language--null hypothesis, test statistic, P-value. Another reasonable strategy is just to teach the idea, and skip the language; however, instructors will have to do some work to use the text that way.

We decided to focus on one test first, developing the ideas and the language in that case, and only then moving on to other tests and generalizations about them. We chose to start with the z-test, as it is relatively simple and widely used. One-tailed tests are used throughout this chapter and the next, as students find them more natural than the two-tailed variety. (There are enough other complications to justify postponing this one to section 29.2.)

Section 1 introduces the idea of the z-test in a tax example. The example has a special feature, that on the null hypothesis, the average of the box is $0 rather than some other number. This has both advantages and disadvantages. (In effect, we've done a z-test for paired comparisons, by taking differences.)

Section 2 introduces the null and alternative hypotheses. We used to tell the usual Berkeley story about statistical hypotheses--the prose equivalent to H_0 and H_1. However, the students couldn't make much sense out of it. So we switched to a more primitive version: the null hypothesis says that there is no effect, so the difference is due to chance. The alternative hypothesis says that the effect is real. In the tax example of section 1, there is a difference between the observed sample average of $219, and the expected $0. The null hypothesis says that this just reflects chance variation in the random selection of 100 tax forms for the study. The alternative says that tax collections will go down.

As section 2 tells the students, the next step is to translate these hypotheses into statements about a box model. In the tax example, there would be one ticket in the box showing the difference "new tax - old tax" for each tax payer. The data on the sample of 100 forms is like the result of drawing 100 times at random from the box. (Few students will pay attention to this sentence, but it is the key to all that follows.)

The null hypothesis translates into the statement that the average of the box is \$0: the alternative translates to the statement that the average is less than \$0. As the text points out, the box model is needed to make the z-test: it's what defines the chances. (This argument is taken up again in chapter 29.)

The downside: Some students come away with a confused, portmanteau null--

- the difference is due to chance, and
- the data are like the draws, and
- the average of the box is 0.

Such students don't see how the statements are related. Other students learn to write down a box before they start computing, but don't really see how the box connects to the problem--as their work after the box may show.

Section 3 introduces the *test statistic* z and the *observed significance level* or *P-value*:

$$z = \frac{\text{observed} - \text{expected}}{\text{SE}}, \qquad P \approx$$

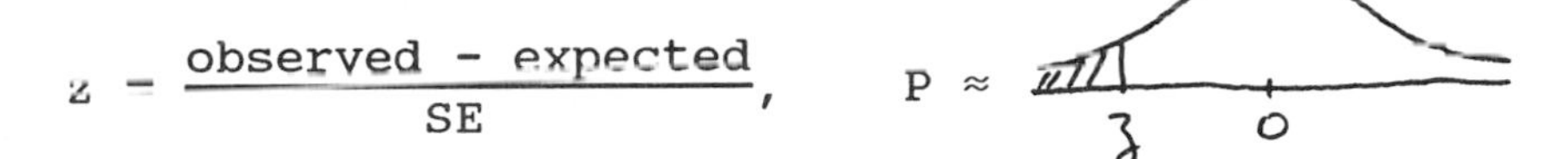

When the P-value of a test is very small, we tend to quote it as a fraction rather than a percent (for instance, on p434). Some students will need help in seeing the connection.

The conventional frequentist interpretation of P is given: if the null hypothesis is right, and the experiment is repeated many times, then P is the proportion of repetitions giving z's more extreme than the observed one. The students are then taught that a test of significance is an argument by contradiction. (Not an easy pitch to make, because many of them don't know what an argument by contradiction is.) Exercises 5–6 in set D try to establish the frequentist interpretation of P.

Section 4 reviews the steps involved in making a test, and introduces the 5% and 1% levels: as we tell the students, a result is *significant* if P is less than 5%, *highly significant* if P is less than 1%. (However, we suggest reporting P instead of just saying how it compares to 5% and 1%; we are not enthusiasts of fixed-level testing.) Many students jump to the conclusion that P represents the chance of the null hypothesis being true; measures are taken to prevent this mistake, both in the text and in exercise 2 on p436 as well as exercise 2 on p437. Other students need to be told, more than once, that small P is bad for the null, big P is good for the null (exercise 1 on p436 and 3 on p437).

Section 5 shows how to make the z-test for qualitative data (counts). The lead example is an ESP experiment done by Charles Tart at U.C. Davis. In this example, and many others, we think there is no natural alternative hypothesis about the box: if a subject has ESP, there is no reason to suppose the successive guesses are independent, so $p>1/2$ isn't a plausible hypothesis--there is no p.

After the first edition of *Statistics* was published, Tart replicated his ESP experiment, and found no effect. He attributed this failure to a change in student attitudes: "In the last year or two, students have become more serious, competitive and achievement-oriented...." The replication is discussed in section 29.4.

Exercises 1–5 in set E (pp440–42) go through testing, step by step; number 9 does the sign test. Exercise 8 is interesting, and there are two ways for students to go off the rails:

(i) using the sample SD instead of the population SD, and
(ii) making a two-sample test, using the two SDs.

Instructors will get to see the second mistake only by having the exercise on a quiz, after doing chapter 27.

Our version of the z-statistic is

$$z = \frac{\text{observed - expected}}{\text{SE}}$$

Many students find this equation a bit cryptic, and do not see how get started using it. We ask, "Well, what is observed?" If the observed value is an average, for example, then they need to compute the expected value for the average--and the SE for the average (p440). We stress that z puts the observed value into standard units.

Section 6 does the t-test; we consider this to be a fairly technical topic, and skip it when pressed for time.

Notes on review exercises. The exercises are designed to emphasize the logical steps involved in making a z-test: formulating hypotheses as statements about a box model, then computing z and P. In many of the exercises--for instance, numbers 8 or 10--students will have a very hard time setting up the box model. Exercise 11 involves computing probabilities by the normal approximation, but some students will try to set it up as a hypothesis-testing problem. Exercise 12 does the sign test.

Note on coverage. We do not introduce the terms *size* or *level*, or use α. The concept of *power* is not introduced: there is enough to do as it is. The connection between tests and confidence intervals is not established: students rarely see the point of isomorphisms.

Chapter 27. More Tests for Averages

Section 1 explains how to calculate the standard error for the difference of two independent chance quantities. Example 2 and exercises 3–4 in set A stress the assumption of independence.

Section 2 presents the two-sample z-test. The context is a decline in math test scores over the period 1973–1982; the tests were administered by the NAEP (National Assessment of Educational Progress). The section shows how to set up the model, with two boxes. Another example does the 0-1 coding. (Our test is the standard one, in disguise; note 3 to the chapter; also see notes 9 and 12.)

Some students will get lost in scaling: for instance, they will figure the difference in percentage points, but its SE in decimals. Exercise 8 on p461 helps. (Ideally, each SE should be seen as the margin of error in some observed value: but a gap between the ideal and the real may be anticipated.)

Section 3 applies the test to experimental data. The issue is as follows. Suppose there are N subjects; n are chosen at random for the treatment group, and m for the control group, with n+m≤N. If n+m is much smaller than N, there are in effect two separate boxes. (For example, see review exercise 8.) If n+m is comparable to N--and n+m=N is the usual case in clinical trials--the treatment and control averages are dependent, and the difference between drawing with or without replacement matters. In principle, then, it is wrong to model the data as independent samples from two large boxes.

Fortunately or otherwise, this fine point has no consequences for the arithmetic: treating the two samples as independent and drawn with replacement will give an excellent approximation to the SE for the difference between the averages. (See notes 9 and 12 to the chapter.)

In example 4 on p462, the calculation is made blindly; the logic is discussed afterward. This is a very difficult passage. Many students will have to work quite hard to see that the sample averages are dependent; and they will be irritated (not unreasonably) to find that the dependence does not matter--for reasons which will seem quite mysterious. Another stumbling block is mechanical: students may not yet have developed a good grip on the procedure for making a two-sample z-test.

Section 4 presents a real example with qualitative data: an experimental test of "rational" decision theory. The material on experiments is new to the second edition.

Notes on review exercises. The point of number 1 is to make the students distinguish between one-sample and two-sample tests. Exercises 2–3 are straightforward two-sample problems; number 2 hints that averages have better power. In exercise 4, the test cannot be done--dependence (see exercise 4 on p468). Review exercises 5–7 are fairly straightforward experimental setups. Number 8, again on experiments, seems quite hard; students either don't see what is being compared to what, or find the comparisons too unnatural to make. In grading this one, we insist on a substantive conclusion--for instance, that people are poor

"I'm not asking for a raise, Sir. I just want to know how you would react if I did."

predictors of their own behavior, but they tend to live up to their predictions about themselves--as one character in the drawing understands very well.

Chapter 28. The Chi-Square Test

Section 1 presents the χ^2-test for goodness of fit, when the model is completely specified. Students have a hard time deciding when to use the χ^2-test and when to use the z-test; some help is given on p481. The text explains how to read the χ^2-table (p479), and says that it is only an approximation. The mathematical underpinnings for the approximation are also discussed in section 1; the main one is a box model, and this is emphasized in the text. Figure 2 (p480) may help: as the number of rolls goes up-- 60, 600, 6000-- the probability histogram will get closer and closer to the smooth curve. The convergence is slow in more senses than one: the figure shows the histogram for 600 rolls. It is still quite bumpy--and it took 8 hours to compute on a SUN-4.

Probability histogram for the null distribution of the χ^2-statistic in 600 rolls of a fair die. (Continues figure 2 in chapter 28.)

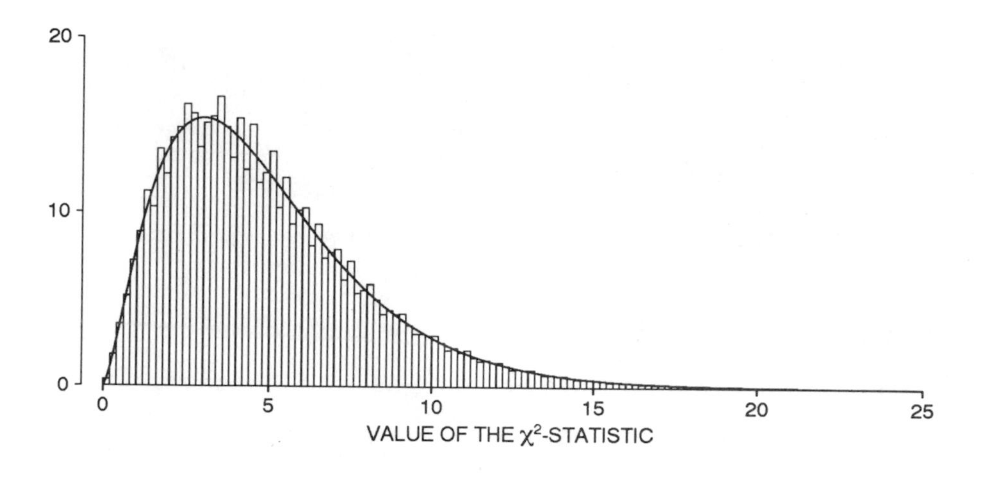

Section 1 closes with a real example--testing the wheel of fortune--new in this edition. Section 2 describes χ^2, in some degree of generality, as a goodness-of-fit test.

Section 3 discusses the pooling of independent χ^2's, and shows how Fisher used the χ^2-test to check up on Mendel (but see note 6 in chapter 25, for the geneticists' counter-arguments). Fisher computed a left-hand tail area, rather than a right-hand tail (p485). Students perceive the choice as a possible trap on a quiz, so this issue will get air time.

Section 4 shows how χ^2 is used to test for dependence in $m \times n$ tables. This will be hard going, but not impossible. (The second edition is quite a bit clearer than the first one.)

Notes on review exercises. Exercises 1 and 4 are straightforward goodness-of-fit questions; number 2 does independence in a 3×3 table. Exercise 3 is a little ambiguous, but left-hand tail areas seem called for. Exercises 5–6–7 are qualitative.

Chapter 29. A Closer Look at Tests of Significance

Many students find tests of significance both complicated and mysterious. Perhaps as a result, the limitations of the technique are often ignored. This can create unnecessary confusion. So we think it is important to discuss what tests of significance don't do. That is the topic of chapter 29.

Section 1 is about fixed-level testing (a procedure we do not recommend).

Section 2 covers data snooping. Students find it very hard to understand that significance levels are compromised by multiple looks at the data. Exercise 6 on pp437–38 and exercise 1 on p498 should help, a little. Exercises 2–5 and 9 in set B (pp498–500) give some practical examples. Students like a definite rule for deciding whether to use a one-tailed or a two-tailed test (pp495–98)--they see exams coming up. This comparatively minor issue will get a lot of attention. (Of course, the application to cholesterol is far from minor: example 2 on p497.)

Section 3 tries to explain that small differences can be statistically significant--or big differences insignificant--depending on the sample size. This point is hard, and irritating. Students have invested a lot of time learning how to operate the tool, they want it to be useful.

Sections 4, 5, and 6 are about the role of the model in testing. Since the arithmetic of the test seems to generate the chances--the P-value--these sections are quite subtle.

Notes on review exercises. Numbers 1 and 2 are straightforward questions, which can be answered from the reading; number 3 is a math question, but a little tricky. Exercise 4 is about data snooping, among other things; hard. Exercise 7 is about not doing tests when you have all the data, an idea the students pick up. Exercises 8 and 9 are on sample design, and are hard. Exercises 6, 10, 11, and 12 are about real studies and raise real questions; very hard.

ANSWERS TO REVIEW EXERCISES

PART I. DESIGN OF EXPERIMENTS

Chapter 2. Observational Studies

1. False. The population got bigger too. You need to look at the number of murders relative to total population size. The population in 1985 was about 240 million, and in 1970 it was about 205 million: 19,893 out of 240 million is just about the same as 16,848 out of 205 million.

2. (a) False. There were a lot more DeVilles on the street. You need to look at rates: 970/157,374 is about 6 per 1000, and 747/68,106 is about 11 per 1000; if anything, thieves prefer New Yorkers.

   (b) False. Taking percents adjusts for the difference in production figures, and that's why statisticians compare percents.

3. No. In the Salk trial, the parents who consented were better off than the parents who did not consent, and their children were more at risk to begin with (p4).

4. Ullyot was using historical controls, which is not such a good idea; the randomized controlled trials were much less positive about the value of the surgery (p8).

5. No. The teachers might have been tempted to put the poorer children into the treatment group. (In fact, this seems to have happened; children in the treatment group were significantly smaller in physical size than the controls, which probably biased the study against the treatment.)

6. (a) They were controlling for possible confounding (p12).

   (b) This is the wrong conclusion to draw. Ex-smokers are a self-selected group, and many people give up smoking because they are sick. So recent ex-smokers include a lot of sick people. (Other epidemiological data suggest that within 5 to 15 years after quitting, ex-smokers are about as healthy as the never-smokers.)

7. No. The data from the double-blind study are more reliable, and suggest that the results from the single-blind were biased.

8. Subjects who did not improve during the first part of the trial probably concluded that they were on the placebo (whether they were or they weren't) and would be switched to the "real" medication during the second part of the trial. This expectation made them improve--the placebo effect.

9. No. In badly designed studies, healthier patients are more likely to be picked for the operation, leaving sicker patients to be the controls. Being a control in a badly designed study may be a marker of illness, but it is unlikely to be a cause of illness. See section 1.2.

10. This is an observational study, so confounding may be a problem. Pill users are more active sexually than non-users, and have more partners. That seems to be what makes the rate of cervical cancer higher among pill users. (This is like ex 10 on p21.)

11. No, the argument is not good. Delinquents tend to come from large families, where most of the children are middle children anyway. For instance, with six children, four out of six have to be middle children. The study should have controlled for family size.

12. False; the conclusion does not follow. This is just like the admissions study (pp16ff). The Democrats may be concentrated in wards with low turnouts. Here is an example, with only two wards (and see ex 13 on p21).

	DEMOCRATS		*REPUBLICANS*	
	Total number	*Number voting*	*Total number*	*Number voting*
Ward A	1000	100	100	5
Ward B	100	60	1000	500

PART II. DESCRIPTIVE STATISTICS

Chapter 3. The Histogram

1. 66 inches, 72 inches.

2. (a) 50% (b) 2%

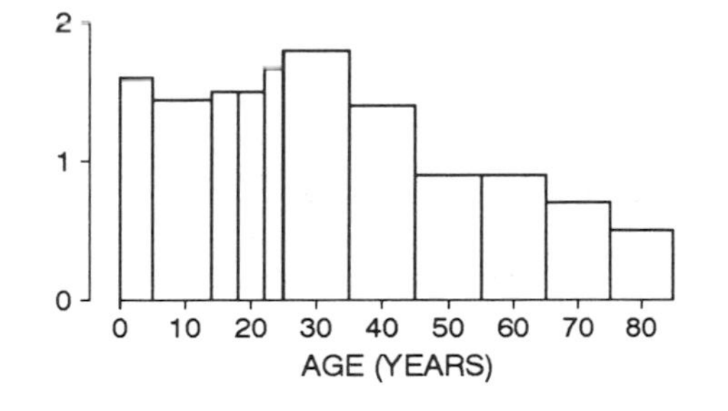

3. (a)

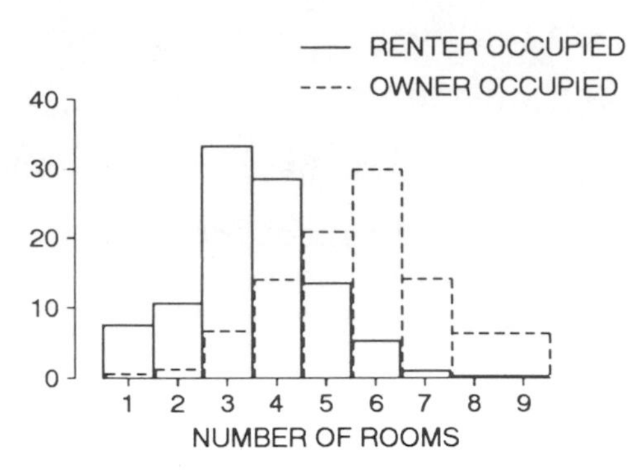

(b) Rounding.

(c) No: taking percentages adjusts for the difference between the total numbers. On the whole, rental units tend to be smaller.

4. (a) 25% (b) 99% (c) 140–150 mm
(d) 135–140 mm (e) About 5×2.1 = 10.5% (f) 102–103 mm
(g) 117–118 mm is a good guess; the interval is somewhere between 115 and 120 mm.

5. \$10 thousand × 1% per thousand dollars = 10%.

6. (a) is right; (b) has a total area of around 200%, and (c) has the wrong units on the density scale.

7. (i) and (ii) not (iii). Reason: With lists (i) and (ii), 25% of the people have heights between 66.5 inches and 67.5 inches; 50% between 67.5 and 68.5 inches; 25% between 68.5 and 69.5 inches. Not so with list (iii).

8. People with failing GPAs may round them up; and 2 is such an important number--for GPAs--that people with GPAs just above 2 may round them down.

9. (a)

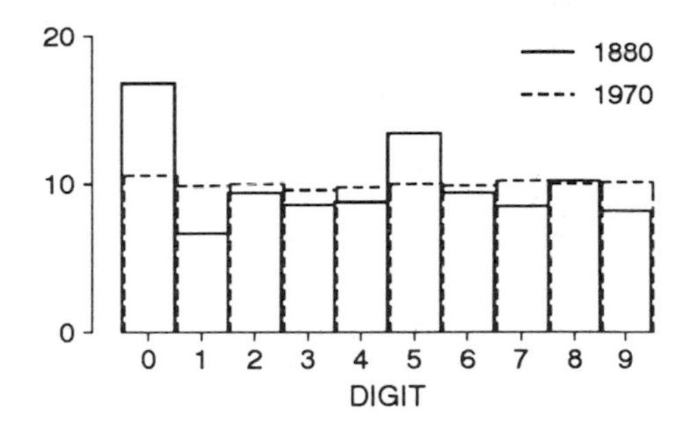

(b) In 1880, people did not know their ages at all accurately, and rounded off.

(c) In 1970, people knew when they were born.

(d) In 1880, there was a strong preference for even digits (although 4 and 6 lose out to 5); again, this is probably due to rounding. In 1970, the preference was much weaker.

10. The lowest of the top 15 scores is 90. Then there is a gap of 6 points--the next score is 84. There is no gap anything like this big in the rest of the distribution.

11. No. There are very few days where the temperature is above 90°; the investigators should have looked at the number of riots divided by the number of days in each temperature range.

Chapter 4. The Average and the Standard Deviation

1. The average is 6; the deviations are 6, -5, -3, 5, 0, -3; the SD is

$$\sqrt{\frac{6^2+(-5)^2+(-3)^2+5^2+0^2+(-3)^2}{6}} = \sqrt{17.3} \approx 4.2$$

2. (a) List (ii) has the smaller SD, because it has more entries at the average.

   (b) This time, list (i) has the smaller SD; list (ii) has two wild entries, 1 and 99.

3. (a) 5: only three of the numbers are smaller than 1, and none are bigger than 10.

   (b) 3: if the SD is 1, the entries 0.6 and 9.9 are much too far from average; the SD can't be 6, because none of the numbers are more than 6 away from the average.

4. (a) True.

   (b) False: all the deviations from average stay the same.

   (c) True.

   (d) True: all the deviations from average are doubled.

   (e) True.

   (f) False: all the deviations from average have their signs changed, but that goes away in the squaring. The SD has to be positive (or, exceptionally, zero).

5. (a) (i) 60 (ii) 50 (iii) 40

   (b) 15: most of the area is within 50 of the average, so 50 is too big; only a little of the area is within 5 of the average, so 5 is too small.

   (c) False. Histograms (i) and (iii) are almost mirror images, and have just about the same SD.

6. (a) average weight of men = 66×2.2 ≈ 145 pounds,
SD ≈ 20 pounds;
average weight of women = 121 pounds,
SD ≈ 20 pounds.

(b) 68%: the range is average ± 1 SD.

(c) bigger than 9 kg: if you take the men and the women together, the spread in weights goes up.

7. (a) The girls have to be taller than the boys at age 11, to bring the average up from 146 cm to 147 cm.

(b) (136+146)/2 = 141 cm.

8. For income, the average will be bigger than the median--long right hand tail at work (p30). For education, the average will be smaller than the median; the histogram has a long left hand tail (p37). And see pp58–59.

9. The SD is about 1 year. Reason: The average is about 18 years. If the SD is 1 month, all the freshmen are about 18, give or take a month or so; that can't be. If the SD is 5 years, there would be a lot of 13-year-olds and 23-year-olds in the freshman class.

10. False: this study is cross-sectional, not longitudinal.

Comment. Longitudinal studies show that real incomes increase roughly linearly with age, until well past 50. The rate of increase goes up for successive birth cohorts, and so do starting salaries. That produces the curvilinear pattern in cross-sectional data. A reference is *The Economic Report of the President*, 1974, pp147ff.

11. (a) 33

(b) $5: the SD is the r.m.s. deviation from average.

12. \$5

13. 80 mm is low--nearly 2 SDs down.

    115 mm is about average--less than an SD down.

    135 mm is about average--less than an SD up.

    210 mm is high--by 4 SDs.

14. This is false, because the data are cross-sectional not longitudinal. For example, it is likely that the people who drink, smoke, and don't eat breakfast die faster.

Chapter 5. The Normal Approximation for Data

1. (a) 79% of 25 ≈ 20

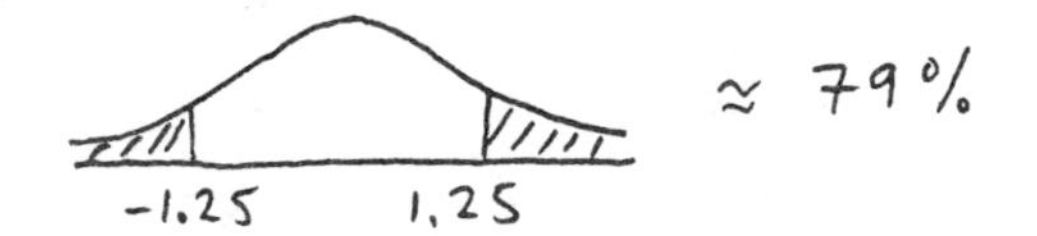

   (b) 18

2. Yes. The entries should be around 1 in size, and they are way too big. The first entry, for example, is 6.2 SDs below average; the eighth entry is even farther down.

Technical comment. By Chebychev's inequality, it is mathematically impossible for a list of 100 numbers to have an entry more than $\sqrt{99} \approx 10$ SDs away from average.

3. (a) Yes: the average goes up by (\$986,000-\$98,600)/100 = \$8,874.

   (b) No. (And that is one advantage of the median--it is not thrown off by outliers.)

4. (a) In 1967, a score of 700 was about 2.10 SDs above average.

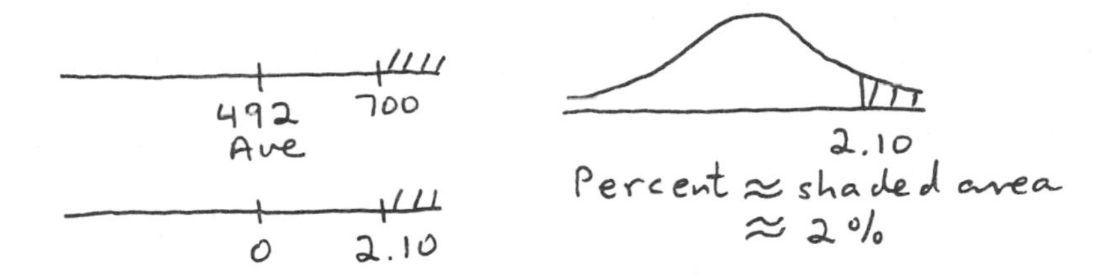

(b) In 1987, a score of 700 was about 2.25 SDs above average, and the percentage is about 1%.

5. (a) 16% (b) 7%

6. (a) 500 (b) 38%

7. No. For example, the normal curve says that about 16% of the scores should be more than 1 SD above average, and none are: the highest possible score is 48.

8. The estimate would be quite a bit too high. The correct percentage, from the histogram, is about 3%; the estimate is 11%.

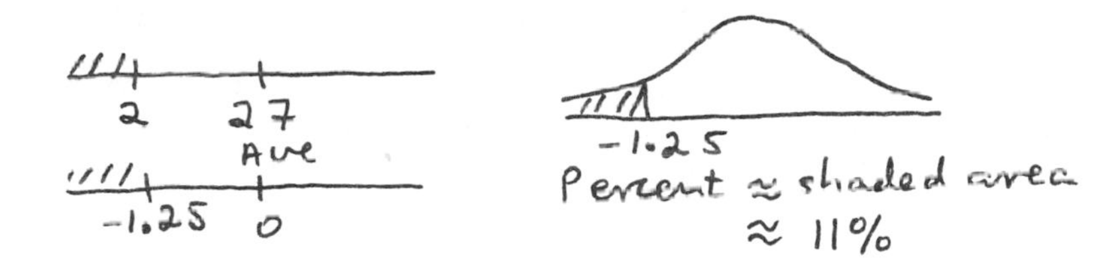

9. If each household consisted of one or more families living together, average household income would be larger than average family income. However, some households consist of just one person living alone. These persons tend to have low incomes, and pull the average down.

10. The percentage of men with heights between 66 inches and 72 inches is exactly equal to the area between 66 inches and 72 inches under the histogram. This percentage is approximately equal to the area between -1 and 1 under the normal curve.

11. (a) False. For example, the list 1,2,99 has a median of 2 and an average of 34.
    (b) False. For the list 1,2,99, two-thirds of the entries are below average.
    (c) False. For example, income data or weight data have a long right-hand tail; educational levels have a long left-hand tail as well as bumps at 8, 12 and 16 years.
    (d) False. If the histogram for a list follows the normal curve, the percentage of entries within 1 SD of average will be around 68%. Otherwise, the percentage could be a lot higher. For instance, take a list of eight numbers: one is 30, one is 70, and six are 50. The average is 50, the SD is 10, and 75% are within 1 SD of average.

12. Histogram (i) is right. With (ii), the average of 1.1 would be right in the middle, and a lot of people would be taking fewer than 1.1 - 1.5 = -.4 courses. Histogram (iii) is even worse.

Chapter 6. Measurement Error

1. False: each measurement is thrown off by chance error, and this changes from measurement to measurement. (It is a good idea to replicate measurements, so as to judge the likely size of the chance error.)

2. (a) The tape may have stretched; the hook at the end may have moved.

   (b) Cloth.

   (c) Yes, as it stretches or the hook moves.

3. (a) False: chance errors are sometimes positive and sometimes negative, but bias goes in one direction only.

   (b) False, same reason.

   (c) True.

4. Probably, a recording error was made at one interview or the other.

5. 0.03 inches or so.

6. (a) No. Persons 2 and 10 copied from each other: they got exactly the same answers, with the decimal point in the wrong place. The other students seem to have worked independently.

   (b) First, nobody got the same answer both times. Second, there is a lot of person-to-person variation.

7. (a) All the numbers on the list are the same. Example: 2,2,2.

   (b) All the numbers on the list are 0.

8.

35 50 70
Ave

-1.5 0 2

-1.5 2

Percent ≈ shaded area
≈ 91%

9. You should guess the average, which is 50; the chance of winning is about 68%.

10. False; see pp82 and 84.

11. (a) 69×1/2 + 63×1/2 = 66 inches.
    (b) Somewhat more than 3 inches: putting the men and women together makes the spread bigger.

12. (a) 69×2/3 + 63×1/3 = 67 inches.

    (b) Same answer as for 11(b).

13. There were a lot more judges in 1988 than in 1789.

14. This is an observational study, so there may be some confounding: drivers may be more at risk than conductors to start with. (In fact, London Transport issued uniforms to drivers and conductors when they were hired, and a record was kept of the sizes; as it turned out, the drivers were heavier than the conductors at time of hire.)

15. The experimental comparison is treatment (groups B + C) versus controls (group A). The investigator is making observational comparisons, because treatment cases select themselves into group B or C. Cases that opt for a pretrial conference are probably different from cases that don't--so there will be lots of confounding. Similar issues came up when comparing the consent and no-consent groups in the Salk vaccine field trial (section 1.1), or the adherers and the non-adherers in the clofibrate trial (section 2.2), or the examined and refused groups in the HIP trial (ex 8 on pp20–21). [Answer continues on next page.]

15. (continued)

    Furthermore, the investigator does not seem to be presenting all the data. For example, there were 2954 cases in all; but only 2780 are reported in the first table. Also, 22% of 701 ≈ 154 of the group B cases reached trial, according to that first table; but the second table only reports on 63 cases in group B.

16. The probable explanation is digit preference; people round their incomes to the nearest $1,000 or $10,000. The class interval $10,000–$12,500 includes the left endpoint (a beautiful round number) as well as $11,000 and $12,000. The next class interval, $12,500–$15,000 includes $13,000 and $14,000--but not $15,000, because class intervals include left endpoints but not right endpoints. So this second interval has only two round numbers in it, rather than three; and neither is as round as $10,000. The other pairs of intervals are similar to these two.

17. No. The 20-year-olds were born 1956–60; the 70-year-olds, 1906–10. In the early part of the century, there was much more pressure to conform and be right-handed.

Comment. Some investigators believe that left-handed people suffer higher mortality; see Stanley Coren, "Left-handedness and accident-related injury risk" *American Journal of Public Health* vol 79 (1989) pp1040–41: however, his evidence seems a little shaky.

PART III. CORRELATION AND REGRESSION

Chapter 8. Correlation

1. The answer is (d). With (a), the averages are too low. With (b), the SDs are too small. With (c), the SDs are too big and the correlation is too high.

2.

0.62	-1.00
-0.85	0.97
0.06	-0.38

3. Option (i) is right: if you take narrow chimneys over 75 and 125 in the scatter diagram, there is more vertical spread--therefore, more uncertainty in the predictions--with the one over 125.

4. (i) 0.60 (ii) 0.30 (iii) 0.95
   Reasoning: correlation (iii) must be nearly 1; correlations (i) and (ii) are moderate, with (i) being stronger.

5. (a) Negative: older cars get fewer miles per gallon.

   (b) Richer people own newer cars, and maintain them better. (Some rich people own Ferrari gas guzzlers, but not many; and 10-year-old Chevrolets in poor repair guzzle even more.)

6. The correlation would be 1.00: all the points on a scatter diagram (for height of wife vs. height of husband) would lie on a straight line which slopes up.

7. 0.3: Taller men do marry taller women, on average; but there is lots of variation around the line.

8. r = -1: no. wrong = 10 - no. right, so all the points on a scatter diagram (for no. wrong vs. no. right) lie on a straight line which slopes down.

9. (a) -0.80 (b) 0.3 (c) 1.00

Comment. In (c), all the points lie on the line y = 2x, so there is no need to do any arithmetic.

10. This is false (p120).

11. (a) 42 inches (b) 2.5 inches (c) 0.80 (d) solid

12. (a) Three students got 91 on the first count and 82 on the second, so they probably worked together. Another three got 85 on both counts; however, since that is the right answer, these students are probably just good counters.

(b) False.

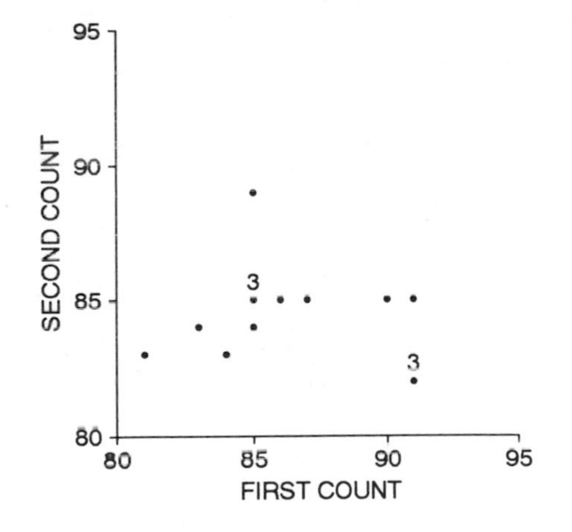

Chapter 9. More About Correlation

1. Histogram, scatter diagram.

2. (a) False. The scatter diagram slopes down: if x is below average, y is generally above average.

[The answer to 2(b) is on the next page.]

2. (b) False. For example, height at age 75 is usually less than height at age 18, but the correlation is positive.

3. (a) Height at 16 and 18: it is easier to predict 2 years ahead than 14.

   (b) Height: environment, personality etc. affect weight more than height, and introduce more variability around the line.

   (c) Age 4: by age 18, these other factors have had more time to introduce variability.

4. Somewhat higher; see exercise set B.

5. Negative: as chlorophyll concentration becomes bigger, the water gets murkier, and Secchi depth becomes smaller.

6. (a) 7: the line $y = 2x-1$ goes through (1,1) and (2,3), then through (4,7).

   (b) Not possible: (1,1), (2,3), and (3,4) do not lie on a line.

7. No: data set (ii) is obtained by adding 3 to the y-values in data set (i), so r has to be the same for both.

8. No: section 4.

9. False: the data are cross-sectional, not longitudinal. Younger people were born later and are better educated, because educational levels have been going up over time.

10. (a) False: the two sections (C and I) that liked their TA best did worst.

(b) True: there is no pattern in the scatter diagram.

(c) False: there is a moderate, positive association (r≈0.5)

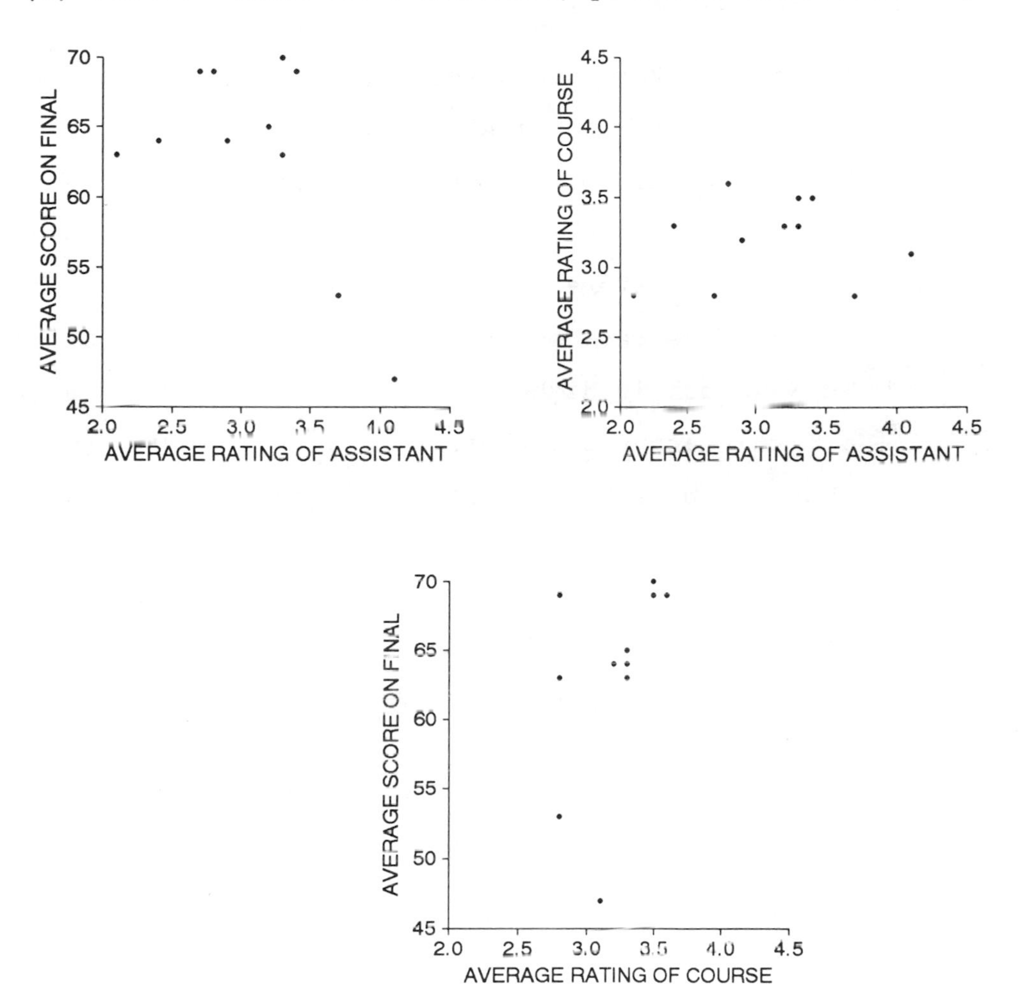

11. No, the conclusion is not justified. Persons who actually protest could find out that the KGB is inefficient. Or, people with a certain kind of personality may be more likely to say both that they participated in political protests and that the KGB is inefficient.

12. (a) True.
    (b) True: the correlation between y and x equals the correlation between x and y.
    (c) False, as (b) should demonstrate.
    (d) False.

13. (a) The test-takers are a self-selected group, and good students are more likely to take the test. As test-takers get to be a larger and larger percentage of high school graduates, the average skill level goes down.
    (b) False: the average in New York is lower, but that seems to be because the percentage who take the test in New York is higher. (Actually, both states are a little above the regression line, and New York is above by more than Wyoming! See note 16 to chapter 9.)

Chapter 10. Regression

1. (a) 63 inches, the average (b) same
   (c) 64 inches (d) 62 inches

   Work for (c): The husband is 4 inches, or $4/2.7 \approx 1.5$ SDs above average in height. The wife is predicted to be above average in height by $r \times 1.5 = 0.25 \times 1.5 \approx 0.4$ SDs. This is $0.4 \times 2.5 = 1$ inch.

2. (a) 112. Work: 115 is 1 SD above average at age 18, so the estimate at age 35 is above average by r SDs. This is $0.8 \times 15 = 12$ points.
   (b) 112.

Comment. The arithmetic is the same for both parts of the question, but the interpretation is different (section 3).

3. (a) 15 years (b) 13.5 years

   (c) Appearances are deceiving; all that is going on here is the regression effect (section 5).

4. So far, it looks like the regression effect (section 4).

5. The regression effect cannot explain a change in the average for the whole study population; these data suggest that patients are more relaxed on the second reading.

6. (a) 50% (b) 50% (c) 21% (d) 65%

   Work for (c). This student was 1.65 SDs below average on the midterm--

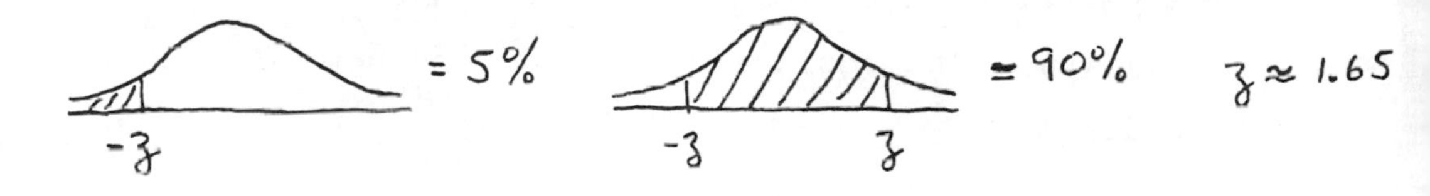

   He should be $r \times 1.65 = 0.5 \times 1.65 \approx 0.8$ SDs below average on the final. Now for the percentile rank on the final--

7. False. This person is likely to be between the 40th and 50th percentiles--regression effect (sections 3-4).

8. (a) False (unless there is something special about the SDs and the point of averages).

   (b) False: r measures association not causation.

   (c) True.

   (d) True: the correlation between y and x equals the correlation between x and y.

   (e) False: r measures association not causation.

9. _ _ _ _ _ y on x

.... SD line

———— x on y

See section 5.

Chapter 11. The R.M.S. Error For Regression

1. (v): see p174.

2. (i) (b). See section 4.

(ii) (a). The residuals show no trend or pattern, they are centered at 0, and they're around $1,000 in size.

(iii) (c). The residuals show a trend upwards. See section 3.

3. Something is wrong. GPAs run from 0 to 4, so if you predict 2, the maximum error is 2; and the computer should be doing better than that.

4. (a) $\sqrt{1-.8^2} \times 2.5 = 1.5$ inches.

(b) $\sqrt{1-.8^2} \times 1.7 \approx 1.0$ inches.

5. (a) r.m.s. error = $\sqrt{1-r^2} \times$ SD of final scores = 12.

(b) 62.2: this student is 20/25 = 0.8 SDs above average on the midterm, and should be above average on the final by $r \times .8 = .48$ SDs; that is 7.2 points.

(c) 12: see part (a).

6. (a)

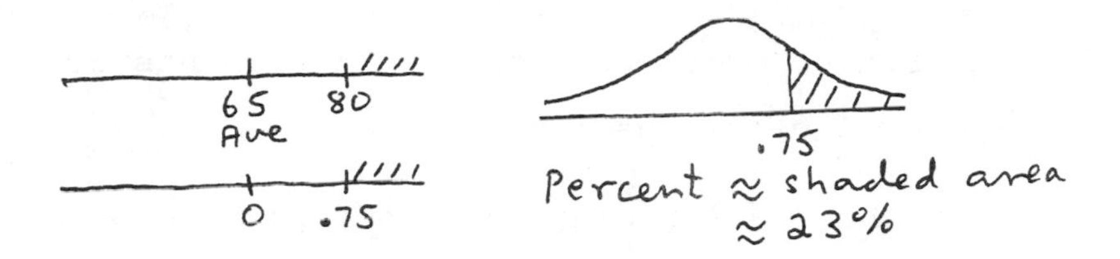

(b) new average ≈ 78, new SD ≈ 17, (80-78)/17 ≈ 0.10

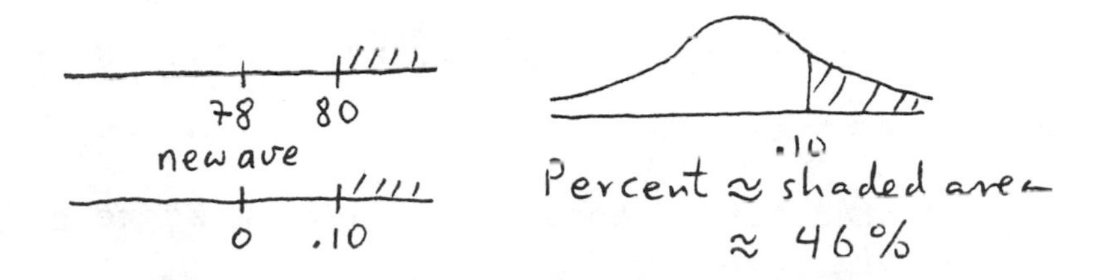

7. Option (iii) is right: regression effect.

8. No. The conclusion seems right, but does not follow from the data. It could be, for example, that better students spend more time doing homework anyway.

9. (a) The procedure is subject to error; replication gives more accuracy, and also lets you gauge the error.

   (b) Students usually prefer not to base their grade on one examination only. Another example is getting a second opinion before surgery.

10. Could be, but this looks a lot like the regression effect.

11. There are a lot of people with 9, 13, and 17 years of education; very few with 8, 12 or 16: and nobody with 0, although three people have 1 year of education. It looks as if 1 year got added to the educational levels, by mistake.

Chapter 12. The Regression Line

1. (a) The r.m.s. error is about 1: the points are 1 or so above or below the line.

   (b) No: the scatter diagram slopes down, and so would the regression line.

2. In a run of one SD, the regression line rises r×SD. The slope is 0.60×20/10 = 1.2 final points per midterm point. The intercept is 55 - 1.2×70 = -29. So the equation is

   predicted final score = 1.2×midterm score - 29.

3. The equation is

   predicted length = 0.05×load + 439.01 cm.

   (a) 439.06 cm (b) 439.26 cm (c) 439.46 cm

   (d) Not possible. This value is too far away from the data. For example, the wire might break under this load.

4. "3 pounds per inch" and "30 pounds per inch" are the only two options with the right units, and the first of these is much too small. The answer is, 30 pounds per inch.

5. The slope is 4 lbs per in ≈ 4 × 0.45 kg per 2.5 cm ≈ 0.72 kg per cm. The intercept is -130 lb ≈ -130×0.45 ≈ -58.5 kg.

6. This is not legitimate (section 10.5). For the regression of height on weight, the slope is

   r×SD of height/SD of weight = 0.40×2.5/25 = 0.04 in/lb.

   The intercept is 62.3 inches.

7. Predicted income is (\$480 per inch)×height - \$18,700. Taller people make more money, on average. Probably, this reflects other variables in family background; although looking every inch an executive may not hurt.

8. Obviously not. The slope means that the 141-pound men are taller, on average, than the 140-pound men--by around 0.047 inches. Similarly, the men who weigh 142 pounds are on average a little taller than the men who weigh 141 pounds. And so forth.

9. This is the regression line of IQ on parental income--in disguise (section 10.2). The slope is

$$r \times \text{SD of IQ}/\text{SD of income} = 0.50 \times 15/\$15{,}000 = 1/\$2{,}000.$$

10. \$41,000 is likely to be too high. You need the other regression line, for income on IQ (section 10.5). The right estimate is \$26,000.

11. No: there doesn't even seem to be any association.

12. This testimony seems wrong. Without doing the experiment--or working very hard at the observational data--you can't be sure what the impact of interventions will be. Associations in the data might be due to confounding.

PART IV. PROBABILITY

Chapter 13. What Are the Chances?

1. (a) False. Chances have to be between 0%--can't happen--and 100%--must happen. See p208.

[The answer to 1(b) is on the next page.]

1. (b) True. The event will happen about 9 times out of 10, and the opposite event will happen the remaining 1 time out of 10. See pp208–9.

2. 100% - 49.8% = 50.2%. See p209.

3. Option (i) is better, because you only have to jump one hurdle, rather than two.

4. There are 42 cards left in the deck, and 4 aces; the chance is $4/42 \approx 10\%$.

5. $4/52 \times 3/51 \times 2/50 \times 1/49 \times 4/48 \approx 1/3{,}000{,}000$.

6. Yes. If the ticket is white, then there are two chances in three to get the number 1, and one chance in three to get 8. And the same for black.

7. (a) True.

   (b) True.

   (c) False: the two events are dependent. See ex 4 on p219.

Comment. The chance of two aces is $4/52 \times 3/51$, not $4/52 \times 4/52$.

8. Both sequences are equally likely, having chance $(1/2)^5$. (Of course, there are a lot more sequences with 3 heads and 3 tails, but that's not the question.)

9. On one roll, the chance is $4/6 \approx 67\%$; on 3 rolls, the chance is $(4/6)^3 \approx 30\%$.

10. The chance of getting four sixes is $(1/6)^4 = 1/1296$. The chance of not getting four sixes is $1 - 1/1296 = 1295/1296 \approx 0.9992$.

11. Box (ii) is better; box (i) has more 1's, and the same 5. With box (ii), you expect to make around $300.

Chapter 14. More about Chance

1. (a) False: these events aren't mutually exclusive, so you can't add the chances. (To find the chance, read section 14.3.)

   (b) False: same reason.

2. Option (i) is better: even if you miss the first time, you get a second try at the money.
3. Option (ii) is better, because there are a lot more ways to win:

   clubs diamonds hearts spades

   or

   hearts diamonds clubs spades

   or

   etc.

4. If you want to find the chance that at least one of the two events will happen, check to see if they are mutually exclusive; if so, you can add the chances.

   If you want to find the chance that both events will happen, check to see if they are independent; if so, you can multiply the chances.

5. (a) True.

   (b) True; see example 2 in chapter 13. (The problem is asking for the unconditional chance--it puts no condition on the first card.)

   (c) False; the events are dependent. (The chance is $4/52 \times 3/51$.)

   (d) False; the events aren't mutually exclusive. (The chance is $4/52 + 4/52 - 4/52 \times 3/51$.)

6. The chance is 6/36 = 1/6, from figure 1 in chapter 14. (Another argument: Imagine one of the dice is white and the other is black; no matter how the white one lands, the black one has 1 chance in 6 to match it.)

7. From figure 1 in chapter 14, the chance is 2/36.

8. This is like the Chevalier de Méré; the chance is $1 - (3/5)^4 \approx 87\%$.

9. 100%--you can't avoid it.

10. Draw a picture like figure 1. There are 12 outcomes, and each has chance 1/12.

1 1	1 2	1 3	1 4
2 1	2 2	2 3	2 4
3 1	3 2	3 3	3 4

For instance, 3 2 means you got 3 from box A and 2 from box B.

(a) The good outcomes are 2 1, 3 1, 3 2; the chance is 3/12 = 25%.

(b) 3/12 = 25%.

(c) 100% - (25%+25%) = 50%.

11. (a) False. A and B can both happen together, with chance $1/3 \times 1/10$. So they aren't mutually exclusive.

(b) True. If A happens, the chance of B drops to 0. That's an extreme form of dependence.

12. The same. There are 60 rolls, and 60 draws; on each roll--and each draw--you have 2 chances in 6 to win $1 and 4 chances in 6 to get nothing.

13. Option (ii) is better. Each time, you have the same 50% chance of winning $1; but you never have to pay anything back.

Chapter 15. The Binomial Coefficients

1. $\frac{6!}{1!5!}\left(\frac{1}{6}\right)^{1}\left(\frac{5}{6}\right)^{5} \approx 40\%$.

2. The chance of not getting a six in one roll is 5/6, and option (iii) does it by the multiplication rule. (Or use the binomial formula.)

3. The chance of getting 4 girls is $(1/2)^4 = 1/16$. The chance of getting 3 girls is 4/16, by the binomial formula; the total chance is 5/16, by the addition rule.

4. False. The binomial formula does not apply. The draws are dependent, because they are made without replacement.

5. True. The first person gets $\frac{0!}{2!6!} = 28$ committees. (Write the 8 names out in a row; then put down 2 C's and 6 N's under the names, where C means "in the committee" and N means "not in the committee"; the binomial coefficient tells you how many ways there are to do that.) By similar reasoning, the second person gets $\frac{8!}{5!3!} = 56$ committees.

6. False. If you choose two names to be on the committee, you are also choosing six to be left out.

7. The chance of getting 2 heads among the first 5 tosses is

$$\frac{5!}{2!3!}\left(\frac{1}{2}\right)^{5}.$$

The chance of getting 4 heads among the last 5 tosses is

$$\frac{5!}{4!1!}\left(\frac{1}{2}\right)^{5}.$$

The first 5 tosses are independent of the last 5. So the answer is

7. (continued)

$$\frac{5!}{2!3!}\left[\frac{1}{2}\right]^5 \times \frac{5!}{4!1!}\left[\frac{1}{2}\right]^5 = \frac{50}{1024} \approx 5\%.$$

8. You need the chance of getting 7, 8, 9, or 10 heads when a coin is tossed 10 times. Use the binomial formula, and the addition rule:

$$\frac{10!}{7!3!}\left[\frac{1}{2}\right]^{10} + \frac{10!}{8!2!}\left[\frac{1}{2}\right]^{10} + \frac{10!}{9!1!}\left[\frac{1}{2}\right]^{10} + \frac{10!}{10!0!}\left[\frac{1}{2}\right]^{10} = \frac{176}{1024} \approx 17\%.$$

PART V. CHANCE VARIABILITY

Chapter 16. The Law of Averages

1. 100 tosses. As the number of tosses goes up, the number of heads is less and less likely to be exactly equal to half the number of tosses, because there are more and more possible values nearby.

2. Option (iii) is the best: the chance error is not likely to be 0 exactly, but should be small relative to the number of draws.

3. Now option (i) is it.

4. False. The percentage of heads equals 50% just when the number of heads equals 50. (And there is about an 8% chance for this to happen--section 18.4.)

5. Possibility (i) is better. Reason: 10/15 = 20/30 = 2/3. So, it's like tossing a coin 15 or 30 times, and asking for 2/3 or more heads. With the bigger number of tosses, this is less likely.

6. 25 draws from | 4 | -1 | -1 | -1 | -1 |.

7. 50 draws from | 4 [$8]'s 34 [-$1]'s |.

8. Both are wrong: luck and the law of averages have nothing to do with it--the chances stay the same, every time. "*The roulette wheel has neither conscience nor memory.*"

9. (a) 30/200 = 0.15 (b) -0.1 (c) average = sum/200

   (d) 5/200 = 0.025; the options describe the same event in different language, so they're the same.

Chapter 17. The Expected Value and Standard Error

1. (a) 100, 1000.

   (b) The average of the box is 7 and the SD is 3. So the expected value for the sum is 100×7 = 700 and the SE is $\sqrt{100} \times 3$ = 30. The sum will be around 700, give or take 30 or so. The chance is about 90%.

650 700 750
Exp
-1.65 0 1.65

-1.65 1.65
Chance ≈ shaded area
≈ 90%

2. (a) The net gain is the like the sum of 100 draws made at random with replacement from the box

| 12 [$2]'s 26 [-$1]'s |

[Answer continues on next page.]

2. (a) (continued)

The average of the box is $-\$2/38 \approx -\$.05$, and the SD is

$$[\$2-(-\$1)] \times \sqrt{\frac{12}{38} \times \frac{26}{38}} \approx \$1.40$$

The net gain will be around $100\times(-\$.05) = -\$5$, give or take $\sqrt{100}\times\$1.40 = \$14$ or so.

(b) The number of wins is like the sum of 100 draws made at random with replacement from the box

12 [1]'s 26 [0]'s

The number will be around 32, give or take 5 or so.

(c) Both bets pay 2 to 1, but roulette gives you a better chance of winning-- $12/38 \approx 32\%$ compared to 25% for Keno. You lose faster at Keno.

3. (a) (iii), using the short-cut.

(b) (i) (c) (v) (d) (iv) (e) (ii)

4. The number of aces in 180 rolls of a die is like the sum of 180 draws from the box | [1] [0] [0] [0] [0] [0] |. So the number of aces will be around 30, give or take 5 or so. There is about a 99.7% chance that the number of aces will be in the range 15 to 45. About 99.7% of the people should get a number in that range.

5. The larger number is worse; the chance error is likely to be larger (in absolute terms) with the larger number of throws.

6. (a) Option (ii) is right: the sum is equally likely to go up or down 1 on each draw, just like the difference. Option (i) is out, you can't add words; with option (iii), the sum can't go up; with option (iv), the sum can't go down.

(b) Expected value = 0, SE = $\sqrt{100}\times 10$.

7. (a) 52.8% (b) 437

(c) 490 is 49% of 1000 and 510 is 51% of 1000, so both options describe the same event in different language; they're the same.

8. (a) 310/100 = 3.1 (b) 4.3×100 = 430

(c) The average of the draws will be between 3 and 4 when the sum is between 300 and 400. The expected value for the sum is 350, and the SE is $\sqrt{100}\times 1.7 = 17$, so the chance is about 99.7%.

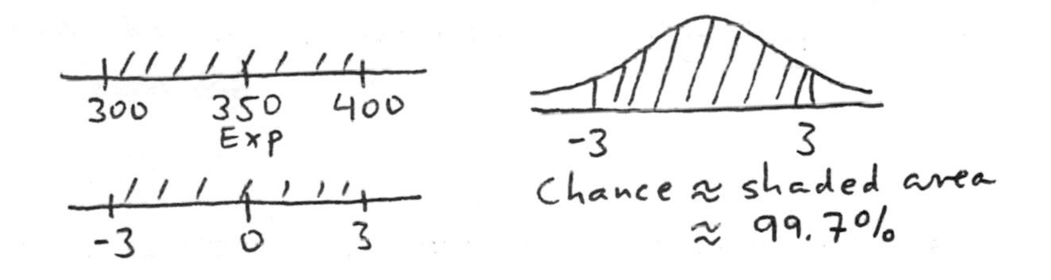

9. (a) is false, (b) and (c) are true. The reason: With A, the net gain is like the sum of 1000 draws from the box

| 12 [$2]'s 26 [-$1]'s |. The average of the the box is -$2/38 and the SD is about $1.39. The expected value for the net gain is -$53 and the SE is $44. With B, the net

9. (continued)

gain has the same expected value, but the SE is bigger; it is $182. Chance variability helps you overcome the negative expected value, so you are more likely to come out ahead with B. You are more likely to win big--or lose big.

The chance of coming out ahead with A is about

The chance of coming out ahead with B is about 38%.

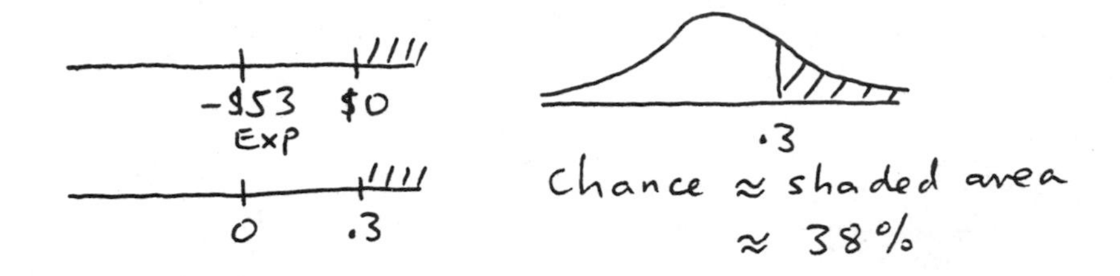

10. If a student answers at random, the score is like the sum of 25 draws from the box | 4 | -1 | -1 | -1 | -1 |. The average of the box is 0, and the SD is 2. So the expected value for the score is 0, and the SE is 10. [Answer continues on next page.]

10. (continued)

(a) The chance of passing is slim to none.

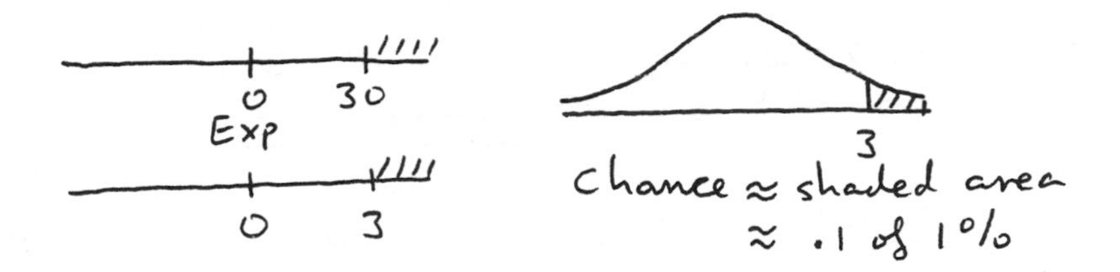

(b) There is about a 16% chance of scoring 10 points or more--

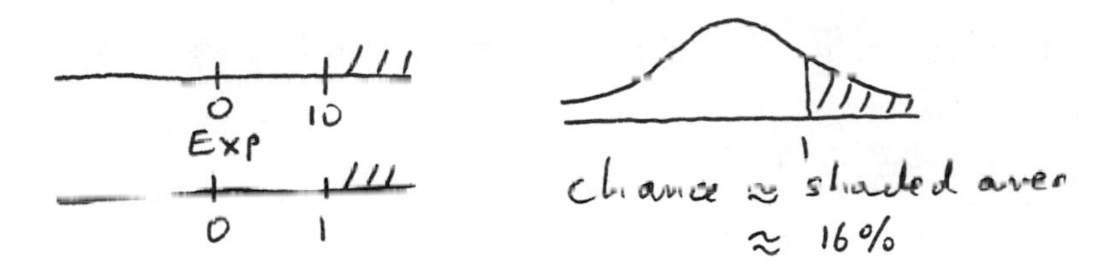

11. The number of heads in each group should be around 50, give or take 5 or so; and the groups are independent. Option (i) is right. With (ii), the number of heads is more like 25. With (iii), there is a strong negative correlation from group to group. And with (iv), the SD is more like 10.

Chapter 18. The Normal Approximation for Probability Histograms

1. 20, 25.

2. (a) The average of the box is 4 and the SD is about 2.24; the expected value for the sum is 1600 and the SE is

2. (a) (continued)

about $\sqrt{400}\times 2.24 \approx 45$.

1500 1600
Exp

-2.25 0

-2.25
chance ≈ shaded area
≈ 99%

(b) The number of 3's is like the sum of 400 draws from the box [0] [1] [0] [0]. The expected number is 100 and the SE is 8.66. The chance is about 12.5%.

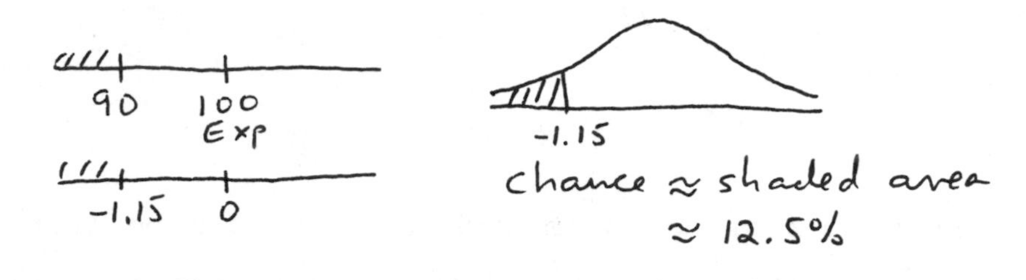

3. The chance that the sum will be in the interval from 10 to 20 inclusive equals the area under <u>the probability histogram</u> between <u>9.5</u> and <u>20.5</u>. The probability histogram gives exact chances, and the normal curve gives approximations--which are often very good.

4. If you get 12 heads, the number of tails is 25-12=13, automatically. The expected number of heads is 12.5, and the SE is 2.5. Use the method of example 1 (p291). The chance is about

4. (continued)

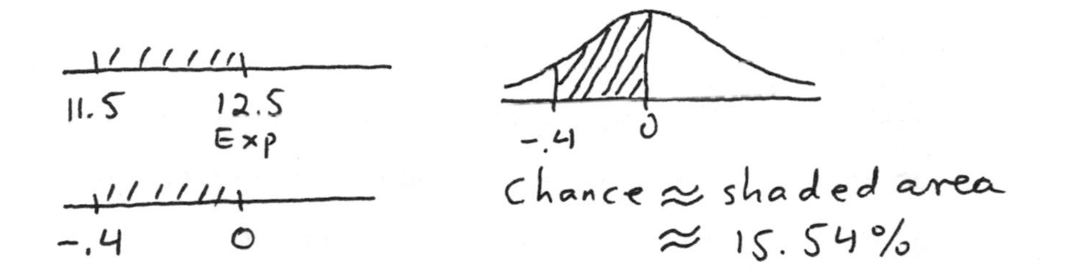

(This problem can be solved using the binomial formula; the exact chance is 15.50%, so the normal approximation looks pretty good.)

5. (i) is the probability histogram for the sum.

   (ii) is the probability histogram for the product.

   (iii) is the histogram for the numbers drawn.

6. Something is wrong with TOSSER. With a million tosses, the number of heads should be around 500,000 give or take 500 or so. TOSSER is 4 SEs too high. Hmmm indeed.

7. (a) Can't be done. For example, the box could have 4 [1]'s and 6 [-10]'s, or, 4 [10]'s and 6 [-1]'s. With these two boxes, the chances would be very different.

   (b) Can be done, using the normal approximation--that's why the average and SD are so useful.

8. (a) This is like drawing at random with replacement from the box [4 [1]'s 6 [0]'s], and asking for the chance that the sum of the draws will be 425 or more. Just use the normal approximation.

   (b) You do not even need the average and SD--see (a).

9. Can't be done. You need to know how many 3's there are in the box.

10. (a) True (p299).

    (b) False. The histogram for the contents of the box does not change, and could have any shape--depending on how the box was set up in the first place.

    (c) False. The histogram for the draws gets more and more like the histogram for the contents of the box. You have to distinguish between the *probability histogram for the sum of the draws*, and the histogram for the draws as data.

    (d) False (p297).

11. Option (ii) is right. With (i), all the values have equal chances, and that's wrong (figure 1, p285).

12. (a) The normal approximation will be too low, because the curve is lower than the probability histogram at 90.

    (b) The approximation will be about right, because the highs and lows cancel.

PART VI. SAMPLING

Chapter 19. Sample Surveys

1. No, because of response bias. The subjects could try to give the interviewers the pleasing answer, rather than the true answer. (In this study, many respondents said they were using the product when they really weren't.)

2. No: this county might be exceptional. (It was.)

3. The people with college degrees were living in more suburban neighborhoods. This was not a good way to draw a sample.

4. This estimate is likely to be too high. With smaller households, the interviewer is less likely to find someone at home. So the survey procedure is, on average, replacing smaller households by larger ones.

5. No. Different sorts of students are more likely or less likely to walk through different parts of the campus at different times, and the chances are difficult (or impossible) to figure. See p313.

6. Option (ii) is better. You win if number of heads is between 480 and 520. That's a broader range than (i).

7. Look at figure 3, chapter 18. The likeliest number of heads is 50: pick that first. Your next two picks should be 49 and 51. Then 48 and 52. And so forth. You should pick 45 through 55. And your chance of winning is about 73%--example 1(a) on p291.

8. (a) 10 through 20.
   (b) 40 through 50.
   (c) To get the center of the range, take 50 from the sum: and then go 5 either way.
   (d) 73%--by exercise 7.

9. This cannot be determined from the information given.

10. The smaller one. As the size of the hospital goes up, the percentage of male births gets closer to 52%, and is less likely to exceed 55%. See section 16.1.

Chapter 20. Chance Errors in Sampling

1. Chance error.

2.

Number of tosses	*NUMBER OF HEADS* *Expected value*	*SE*	*PERCENT OF HEADS* *Expected value*	*SE*
100	50	5	50%	5%
2,500	1,250	25	50%	1%
10,000	5,000	50	50%	0.5 of 1%
1,000,000	500,000	500	50%	0.05 of 1%

3. 10%, 1%. The number of red marbles is like the sum of 900 draws made at random with replacement from the box [[1] 9 [0]'s]. The average of the box is 0.1 and the SD is 0.3. The number of red marbles will be around 90, give or take $\sqrt{900} \times 0.3 = 9$ or so. Convert to percents, relative to 900: the percentage of red marbles among the 900 draws will be 10%, give or take 1% or so.

4. (a) There should be 50,000 tickets in the box--the box is the population (all 50,000 forms).

   (b) Each ticket shows a 0 (gross income under $50,000) or a 1 (gross income over $50,000).

   (c) False: the SD of the box is $\sqrt{0.2 \times 0.8} = 0.4$.

   (d) True: the draws are the sample.

   (e) The number of 1's in 900 draws will be around 180, give or take $\sqrt{900} \times 0.4 = 12$ or so. The percentage of forms in the sample with gross incomes over $50,000 will be 20%, give or take $12/900 \approx 1.33\%$ or so. The chance is about 55%.
   [Answer continues on next page.]

4. (e) (continued)

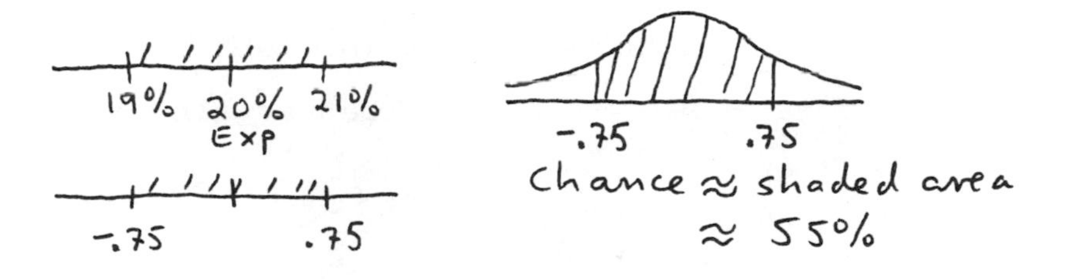

(f) Can't be done with the information given. You need to know the percentage of forms with gross incomes over \$75,000. And you can't use the normal curve, because these data are far from normal.

5. (a) 50,000. (b) A gross income. (c) True. (d) True.

(e) The total gross income is like the sum of 900 draws from the box. The expected value for the total is 900×\$37,000 = \$33,300,000. The SE is $\sqrt{900}\times\$20,000$ = \$600,000. The chance is about 69%.

MM = \$1,000,000

33MM 33.3 MM
Exp
-.5 0

-.5
Chance ≈ shaded area
≈ 69%

In exercise 4, you were classifying and counting, so you had to change the box; here, you are working with a sum, so it would be a bad idea to change the box.

6. The total weight of the group is like the sum of 100 draws from a box; the average of the box is 150 pounds and the SD is 25 pounds. The expected value for the sum is 15,000 pounds and the SE is $\sqrt{100} \times 25 = 250$ pounds. The chance is about 2%--maybe a better elevator is needed!

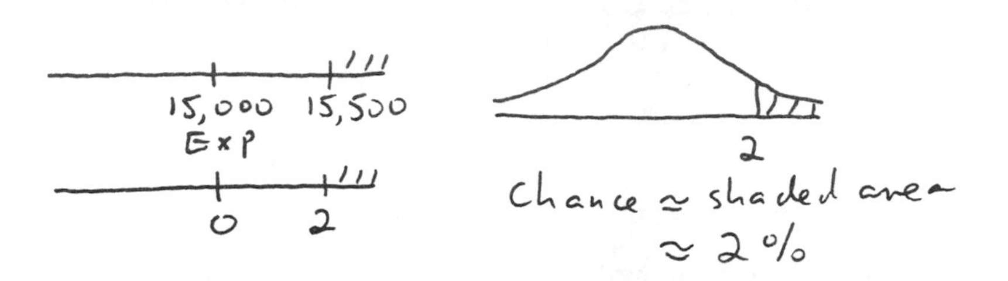

7. Option (ii) is right. In absolute terms, the California sample will be much bigger, and therefore more accurate for estimating percents (section 3).

8. This is fine. Before computing the SE, find out what the SE is for.

9. The total number of interviews is like the sum of 400 draws from a box. The average of the box is 2.38, and the SD is 1.87. The total number of interviews will be around $400 \times 2.38 = 952$, give or take $\sqrt{400} \times 1.87 \approx 37$ or so.

10. The number of sample families without cars is like the sum of 1600 draws from a 0-1 box. There is a ticket in the box for each of the 25,000 families in the town, marked 1 (no car) or 0 (owns cars). So, the fraction of 1's in the box is 0.1, and the SD of the box is $\sqrt{0.1 \times 0.9} = 0.3$. The sum will be around 160, give or take $\sqrt{1600} \times 0.3 = 12$ or so. The percentage of sample

10. (continued)

families without cars will be around 10%, give or take 12/1600 = 0.75 of 1% or so. The chance is about 82%.

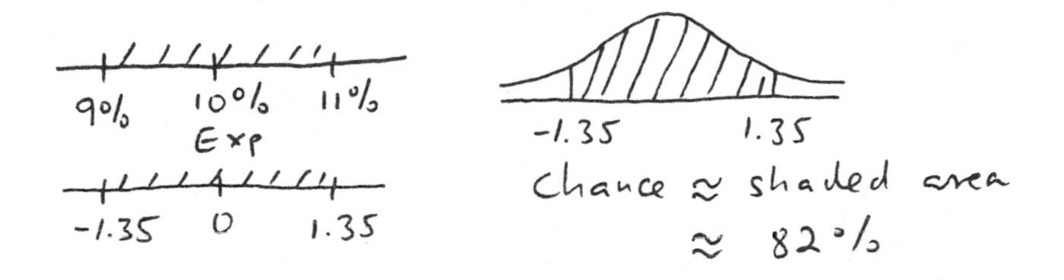

11. The number of blacks in the sample should be around 26, give or take 4.4 or so. The chance of getting 8 or fewer is nearly 0. This was not a random sample, whatever the Supreme Court thought at the time.

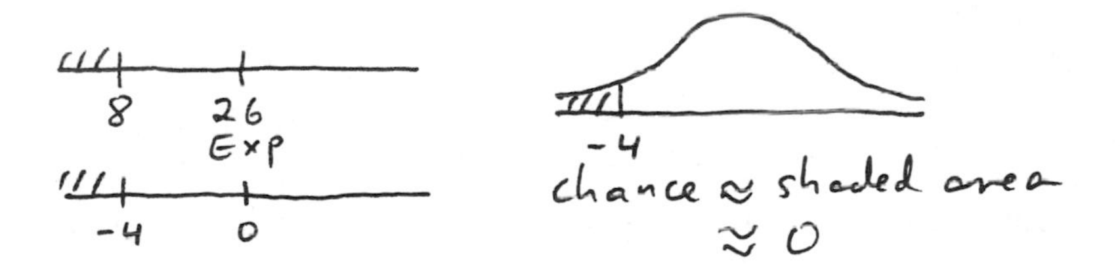

Chapter 21. The Accuracy of Percentages

1. (a) The sample is like 500 draws from a box with 25,000 tickets; each ticket is marked 1 (has a dishwasher) or 0 (does not have a dishwasher). The number of sample households with dishwashers is like the sum of the draws.

1. (a) (continued)

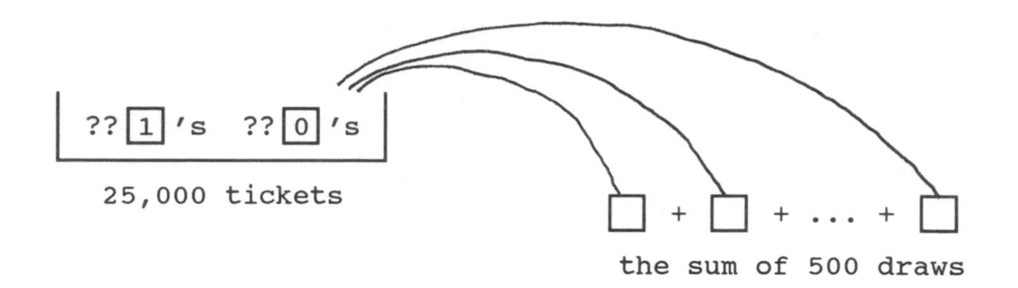

The fraction of 1's in the box is unknown, but can be estimated by the fraction in the sample, as 179/500 = 0.358. On this basis, the SD of the box is estimated as $\sqrt{0.358 \times 0.642} \approx 0.48$. The SE for the number of sample households with dishwashers is estimated as $\sqrt{500} \times 0.48 \approx 11$, and 11/500 = 2.2%. The percentage of households in the town with dishwashers is estimated as 35.8%, give or take 2.2% or so.

(b) The 95%-confidence interval is 35.8% ± 4.4%.

(c) 99.6%, 0.3 of 1%.

(d) Can't be done: the box is so lopsided that the normal approximation won't work. (See ex 5–6 on p350.)

2. In the sample, (172+207)/500 = 379/500 = 75.8% of the households owned cars. The percentage of households in the town with cars is estimated as 75.8%, give or take 1.9% or so.

3. (a) The box has millions of tickets, one for each 17-year-old in school in 1986. Tickets are marked 1 for those who knew that Chaucer wrote *The Canterbury Tales*, and 0 for the others. The data are like 6000 draws from the box, and the number of students in the

3. (a) (continued)

sample who know the answer is like the sum of the draws. The fraction of 1's in the box can be estimated from the sample as 0.361. On this basis, the SD of the box is estimated as $\sqrt{0.361 \times 0.639} \approx 0.48$. The SE for the number of students in the sample who know the answer is estimated as $\sqrt{6000} \times 0.48 \approx 37$. The SE for the percentage is $37/6000 \approx 0.6$ of 1%. The percentage of students in the population who know the answer is estimated as 36.1%, give or take 0.6 of 1% or so. The 95%-confidence interval is 36.1% ± 1.2%.

(b) 95.2% ± 0.6 of 1%.

4. Yes. This is like example 1 on p346. The point is, the estimate is based on a simple random sample.

5. This is not the right SE. We do not have a simple random sample of 252 days, and the daily changes are dependent: each day's closing price is the next day's opening price.

6. This is not the right SE. The bank has mixed up 73 cents with 73%. (The right way to figure the SE for an average will be explained in chapter 23.)

7. False, because of chance error. The sample percentage is likely to be close to the population percentage, but not exactly equal. The SE for the percentage says how far off you can expect to be.

8. Your net gain is like the sum of 100 draws from the box | 6 [$11]'s 94 [-$1]'s |. The average of the box is -$.28, and the SD is about $2.85. Your net gain will be around -$28, give or take $28 or so.

9. 710/100 = 7.1, so the two options describe the same event in different words. They are the same.

10. Option (ii) is better. For example, about 95% of the estimates will be right to within 2 SEs, about 99.7% of them will be right to within 3 SEs, and so forth.

11. (i) is irrelevant, (ii) is a histogram for the numbers drawn, and (iii) is a probability histogram for the sum. Reason: (iii) looks like the normal; (i) would allow 3 and 4 among the draws.

Chapter 22. Measuring Employment and Unemployment

1. (a) False: see section 4.

   (b) They were outside the labor force (section 3).

2. False: this is not a simple random sample (section 5).

3. The estimate is 7.0 million, and the estimated SE is 0.1 million (section 5).

4. Nothing (section 7).

5. (a) Yes: see pp312—13.

   (b) No. For example, the auditors can't get items 17 and 18 in the same sample: with a simple random sample, they could.

Comment. List samples are easier to draw than simple random samples, and for many purposes they work just fine. You have to be a little careful in figuring the SE, though; and it's better to use several random starts than one.

6. False. This is a sample of convenience, not a simple random sample. The formula does not apply (p369).

7. Option (v) is it. The 95%-confidence interval is $49\% \pm 2$ SE, so the SE for the percentage must be $\frac{1}{2} \times 6\%$. The 99.7%-interval is $49\% \pm 3 \text{ SE} = 49\% \pm \frac{3}{2} \times 6\%$.

8. The total of the numbers tossed is like the sum of 200 draws from the box | [1] [2] [3] [4] [5] [6] |. The average of the box is 3.5, so the total should be around 200×3.5 = 700. That's show biz.

9. No, because of response bias. The attitude of the interviewers could well affect the responses. (And in this example, it did.)

10. A simple random sample is drawn at random <u>without</u> replacement (p310).

11. Drawing without replacement gives a more accurate estimate for the percentage of 1's in the box, because you get new information on each draw. (For an extreme example, think about 250 draws.) Also see p337.

Comment. Drawing with replacement leads to easier formulas for the SE. That is why we study it. These formulas are exact for drawing with replacement, and approximate for drawing without replacement. Exercise 11 is asking about the estimate for the percentage of 1's; it is not asking which formula is easier or more accurate.

12. The histogram for the data will get closer and closer to the probability histogram for the number of heads in 100 tosses of a coin (section 18.2). This probability histogram is not the normal curve (chapter 18, figure 3).

Chapter 23. The Accuracy of Averages

1. The sum of 400 draws will be around $400 \times 100 = 40{,}000$, give or take $\sqrt{400} \times 20 = 400$ or so. The average of 400 draws will be around 100, give or take 1 or so.

   (a) The chance is almost 100%.

   (b) The range is "expected value ± 1 SE", so the chance is about 68%.

   Moral: the SE and the SD are different.

2. (a) False (p381).

   (b) True. That's what the SE does for you (p381).

   (c) True. The 68%-confidence interval is

   "sample average ± SE for average".

3. Model: There is a box with 50,000 tickets, one for each each household in the town. The ticket shows the commute distance for the head of household. The data are like 1000 draws from the box. The SD of the box is unknown, but can be estimated by the SD of the data, as 9.0 miles. The SE for the sum of the draws is estimated as $\sqrt{1000} \times 9 \approx$ 285 miles, and the SE for the average is estimated as $285/1000 \approx 0.3$ miles.

   (a) 8.7 miles, 0.3 miles.

   (b) 8.7 miles ± 0.6 miles.

4. Can't be done with the information given, because this is a simple random sample of households not persons. (For example, if a household is far from the center of town,

4. (continued)

all the occupants are likely to have a long commute.) The SE is going to be bigger than the SE for a simple random sample of 1000 persons--although sampling households is generally cost-effective. See section 22.5.

5. Model: There is a box with 50,000 tickets, one for each household in the town. The ticket is marked 1 if the head of household commutes by car; otherwise, 0. The data are like 1000 draws from the box. The fraction of 1's in the box is unknown, but can be estimated by the fraction in the sample, as 0.721. On this basis, the SD of the box is estimated as $\sqrt{0.721 \times 0.279} \approx 0.45$. The SE for the number of 1's in 1000 draws is estimated as $\sqrt{1000} \times 0.45 \approx 14$. The SE for the percentage of 1's is $14/1000 \approx 1.4\%$. The percentage of 1's in the box is estimated as 72.1%, give or take 1.4% or so. The 95%-confidence interval is $72.1\% \pm 2.8\%$.

6. (a) False: 325 is not the SD, and the data aren't normal.

   (b) True: the interval is "average ± SE".

   (c) True; that's what confidence intervals are all about (section 21.3).

   (d) False: the normal curve is being used on the probability histogram for the sample average, not the data (pp382–83).

   (e) False: the data don't follow the normal curve.

7. 1.7 ± 0.1.

8. (a) True.

   (b) False. There is no such thing as a 95%-confidence interval for the sample average, you know the sample average. It's the population average that you have to worry about.

   (c) True.

   (d) False. This confuses the SD with the SE. And it's ridiculous in the first place, because a household must have a whole number of persons (1, or 2, or 3, and so forth). The range 2.16 to 2.44 is impossible for any particular household, let alone 95% of them; although this range is fine, for the average of all the households.

   (e) False. For instance, if household size followed the normal curve, there would be many households with a negative number of occupants; we're not ready for that.

   (f) False. Household size does not follow the normal curve, but you can use the normal curve to approximate the probability histogram for the sample average (pp382–83).

9. (i) is the histogram for the draws.
   (ii) is the probability histogram for the average of the draws.
   (iii) is the histogram for the contents of the box.
   Reason: in (iii), the number of 1's is exactly double the number of 2's; and the number of 4's equals the number of 2's; so this goes with the box, not the draws.

10. This is not a 95%-confidence interval, because the class is a sample of convenience, not a probability sample.

11. You can't estimate the likely size of the chance error.

12. number of draws = 400

    average of box = 3

    expected value for sum of draws = 400×3 = 1200

    sum of draws = 1161

    SE for sum of draws = 40

    SE for average of draws = 40/400 = 0.1

    the SD of the box is $40/\sqrt{400} - 2$

PART VII. CHANCE MODELS

Chapter 24. A Model for Measurement Error

1. Model: each measurement equals the exact elevation, plus a draw from the error box. The tickets in the box average out to 0. Their SD is unknown, but can be estimated by the SD of the data, as 30 inches. The SE for the sum of the 25 measurements is estimated as $\sqrt{25}\times 30$ = 150 inches. The SE for their average is estimated as 150/25 = 6 inches.

   (a) 81,411 inches, 6 inches.

   (b) False. There is no such thing as a confidence interval for the <u>sample average</u>, you know the sample average. It's the elevation of the mountain that you have to worry about.

   (c) True.

[Answer continues on next page.]

1. (continued)

   (d) False: this mixes up SD and SE.

   (e) False, same issue.

   (f) False: see exercise 9 on p385.

2. 299,774 ± 0.6 km/sec.

3. The SD.

4. If they got the length wrong (as they must have, at least by a little), there is a bias in the speed measurements.

5. False. If they're thinking of the average for 1987, they know it. If they're thinking of some other average, there is still a problem, because the Gauss model does not apply.

6. Use the SD of the old data, 18 micrograms (p405). The 95%-interval is 78.1 ± 5.1 micrograms above 1 kilogram.

7. The data are like 100 draws made at random with replacement from a box. The average of the box is about 58 seconds, and the SD is about 2 seconds. The total execution time is like the sum of 100 draws from the box. This will be about $100 \times 58 = 5800$ seconds, give or take $\sqrt{100} \times 2 = 20$ seconds or so.

8. (a) Model: the weights of the sticks are drawn at random from a box; the average of the box is 4 ounces, and the SD is 0.05 ounces. The weight of a package is like the sum of 4 draws. The expected value is $4 \times 4 = 16$ ounces. The SE for the sum is $\sqrt{4} \times 0.05 = 0.1$ ounces. The answer: 16 ounces, 0.1 ounces.
   [Answer continues on next page.]

8. (continued)

(b) The total weight of 100 packages is like the sum of 400 draws from the box. The expected value is

$$400 \times 4 = 1600 \text{ ounces} = 100 \text{ pounds.}$$

The SE for the sum is

$$\sqrt{400} \times 0.05 = 1 \text{ ounce.}$$

The total weight will be 100 pounds, give or take 1 ounce. The range "100 pounds ± 2 ounces" is "expected value ± 2 SE", so the chance is about 95%.

9. (a) True.

(b) False. You can use the variability in the data to estimate the SD of the error box, and then compute a standard error.

10. False. You use the curve on the probability histogram for the average, not the histogram for the data (pp382–83).

11. This procedure invites bias. It would be better to make the measurements some prespecified number of times, and then take the average. Of course, if something went wrong during the measurement process, it might be okay to exclude the result.

Chapter 25. Chance Models in Genetics

1. *First line.* The yellow parent must be y/y. The green parent must be g/y, else, all progeny would be green.

   Second line. Both parents must be g/y, else no yellow progeny.

   Third line. Both parents are y/y.

   Fourth line. The yellow parent is y/y. The green parent must be g/g, else there would be yellow progeny.

   Fifth line. One parent must be g/g, else there would be yellow progeny.

2. Genetic model: one gene-pair, with two variants, smooth (s) dominant and wrinkled (w) recessive. Crossing s/w with s/w produces smooth with probability 3/4, wrinkled with probability 1/4. Mendel had 5474+1850 = 7324 plants. The number of smoothies is like the sum of 7324 draws from the box | 1 1 1 0 |. The expected value is 7324 × 3/4 = 5493. He was off by 5493 − 5474 = 19. The standard error is $\sqrt{7324} \times \sqrt{3/4 \times 1/4} \approx 37$.

no. of smoothies

5474 5493 5512

Exp

−.5 0 .5

−.5 .5

Chance ≈ shaded area ≈ 38%

3. Genetic model: one gene-pair, with two variants, early (e) and late (ℓ). e/e is early, ℓ/ℓ is late and e/ℓ is intermediate. So intermediate×intermediate should give about 25% early, 50% intermediate, 25% late. With 2500 plants from this cross, the number of intermediates is like the sum of 2500 draws from the box [1] [0]. So the expected number is 1250 and the SE is 25. The chance is about 2.5%.

4. (a) No. The man's father contributed the Y-chromosome, and this has nothing to do with baldness.

   (b) Yes. The mother got one X chromosome from her father, and may have passed it on to her son, so he is more likely to go bald.

5. (a) No--the child will have at least one *A*.

   (b) Yes--if both parents are *A*/*a*, and the child gets *a* from each parent.

   (c) No--both parents are *a*/*a*, so the child will be *a*/*a* too.

PART VIII. TESTS OF SIGNIFICANCE

Chapter 26. Tests of Significance

1. (a) True (p435).

   (b) False. The null says it's chance, the alternative says it's real (p432).

2. The data are like 3800 draws made at random with replacement from a box | ?? [0]'s ?? [1]'s |, 1 = red.

   (a) Null: The fraction of 1's in the box is 18/38.

   Alt: The fraction of 1's in the box is more than 18/38.

   (b) The SD of the box (computed using the null) is nearly 0.5, so the SE for the number of reds is $\sqrt{3800} \times 0.5 \approx 31$. The expected number of reds (computed using the null) is 1800. So z = (obs-exp)/SE = (1890-1800)/31 ≈ 2.9, and

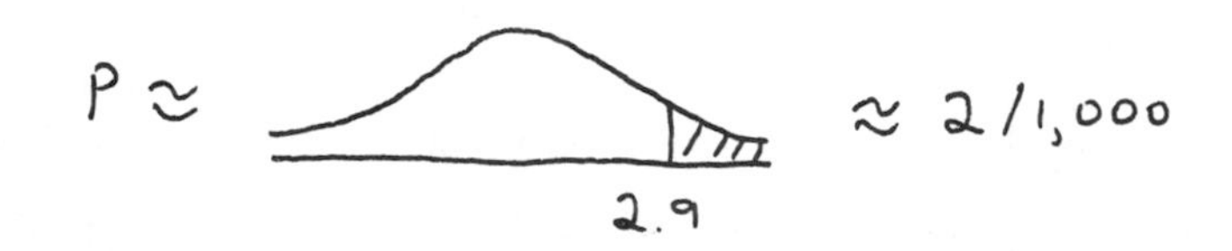

   (c) Yes.

Comments. (i) This problem is about the number of reds. In the formula for z, *obs*, *exp*, and *SE* all refer to the number of reds. The expected, as always, is computed from the null. In this problem, the null gives the composition of the box, so the SD is computed from the null; it is not estimated from the data (p440).

(ii) This problem, and several others below, can be done using one-sided or two-sided tests. The distinction does not matter here; it is discussed in chapter 29.

3. Null hypothesis: data are like 200 draws from the box | [1] [1] [1] [0] |, 1=blue and 0=white. The expected number of blues is 150, and its SE is 6, so z = (obs-exp)/SE = (142-150)/6 ≈ -1.35 and P ≈ 9%. This could be chance.

4. The TA's null: the scores in his section are like 30 draws at random from a box containing all 900 scores. (There is little difference between drawing with or without replacement, because the box is so big.) The null hypothesis specifies the average and the SD of the box:

4. (continued)

62 and 20. The expected value for the average of the draws is 62, and its SE is 3.65. So z = (obs-exp)/SE = (57-62)/3.65 ≈ 1.35, and P ≈ 9%. The TA's defense looks pretty good.

5. The box has one ticket for each freshman at the university, showing how many hours per week that student spends at parties. So there are about 3000 tickets in the box. The data are like 100 draws from the box. The null hypothesis says that the average of the box is 7.5 hours. The alternative says that the average isn't 7.5 hours. The observed value for the sample average is 6.6 hours. The SD of the box is not known, but can be estimated from the data as 9 hours. On this basis, the SE for the sample average is estimated as 0.9 hours. Then z = (obs-exp)/SE ≈ (6.6-7.5)/0.9 = -1. The difference looks like chance.

6. (a) In the best case for Judge Ford, we are tossing a coin 350 times, and asking for the chance of getting 102 heads or fewer. The expected number of heads is 175, and its SE is 9.

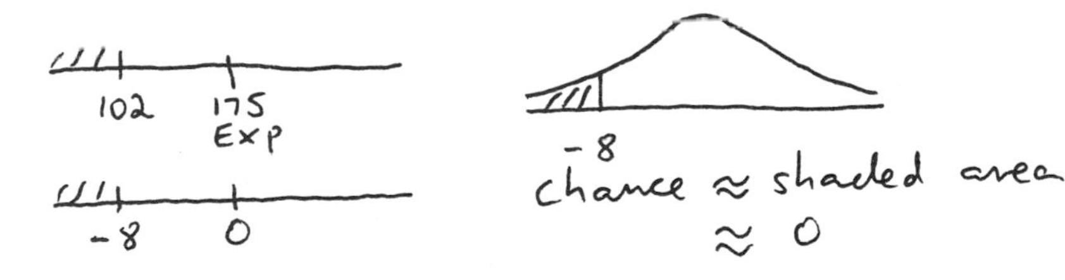

6\. (continued)

(b) 100 draws are made at random without replacement from the box | 102 [1]'s 248 [0]'s |, 1=woman, 0=man. The expected number of 1's is 29. If the draws are made with replacement, the SE for the number of 1's is $\sqrt{100} \times \sqrt{0.29 \times 0.71} \approx 4.54$. The correction factor is $\sqrt{\frac{350-100}{350-1}} \approx 0.846$. The SE for number of 1's, when drawing without replacement, is $0.846 \times 4.54 \approx 4$.

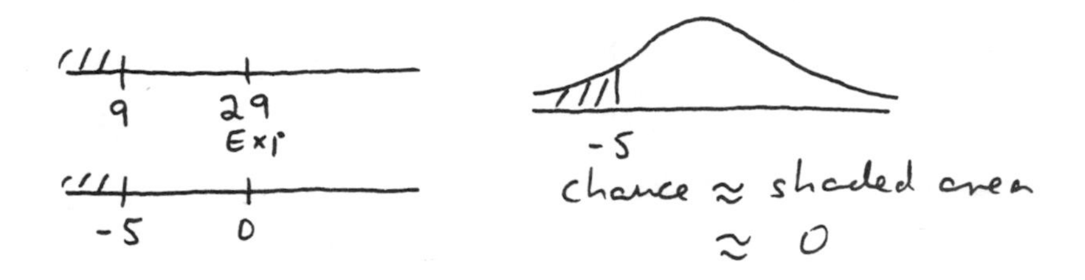

(c) Judge Ford was not choosing at random; he seems to have been excluding women.

Comment. After Hans Zeisel made a statistical analysis of Judge Ford's procedures, the percentage of women jurors went up quite dramatically; see note 14 to chapter 26.

7\. Disagree. There are 580 + 442 = 1022 subjects. With a coin, the expected number in the control group is 511, and the SE is about 16, so the chance of getting 442 or fewer in the control group is practically 0. Both patients and doctors know that if you turn up on an odd day of the month, you get the therapy. There may be a considerable

7. (continued)

temptation to enroll more patients on odd days, and that seems to be what happened.

Comment. The danger is that the patients enrolled on the odd days will be different from the ones enrolled on the even days. For example, the doctors may tend to enroll relatively healthy patients--who need the therapy less--on the even days. If the control group starts off healthier than the treatment group, the study is biased against the treatment. Tossing a coin is safer.

8. Model: There is one ticket in the box for each person in the county, age 18 and over. The ticket shows that person's educational level. The data are like 1000 draws from the box.

   Null: The average of the box is 13 years.

   Alt: The average of the box isn't 13 years.

   The expected value for the average of the draws is 13 years, based on the null. The SD of the box is unknown (there is no reason the spread in the county should equal the spread in the nation), but can be estimated as 5 years--the SD of the data. On this basis, the SE for the sample average is estimated as 0.16 years. The observed value for the sample average is 14 years, so $z = (\text{obs}-\text{exp})/\text{SE} = (14-13)/0.16 \approx 6$, and $P \approx 0$. This is probably a rich, suburban county, where the educational level would be higher than average.

9. Null: the 3 Sunday numbers are like 3 draws made at random (without replacement) from a box containing all 25 numbers in the table. The average of these numbers is 436, and their SD is about 40. The expected value for the average

9. (continued)

is 436, and the SE is 23. The 3 Sunday numbers average 357, so z = (obs-exp)/SE = (357-436)/23 ≈ -3.4, and

Inference: Sundays are for golf.

Comments. (i) Many deliveries are induced, and some are surgical, so obstetricians really can influence the timing. (ii) The 3 Sunday numbers are 344, 377, 351. The number for Saturday, August 6, is also 377, and all the other numbers in the table are larger. In all, there are 25!/(3!×22!) = 2300 samples of size 3. Exactly 2 have an average equal to the Sunday average, and none are smaller. The exact significance probability is 2/2300 ≈ 9/10,000, compared to 3/10,000 from the curve.

10. Model: there is one ticket in the box for each household in the country, marked 1 if the household experienced at least one burglary in 1986, and 0 otherwise. (This is a very large box.) If the FBI data are accurate, the percentage of 1's in the box is 1.3%: this is the null hypothesis. The Survey data are like 50,000 draws made at random from the box. (There is essentially no difference between drawing with or without replacement.)

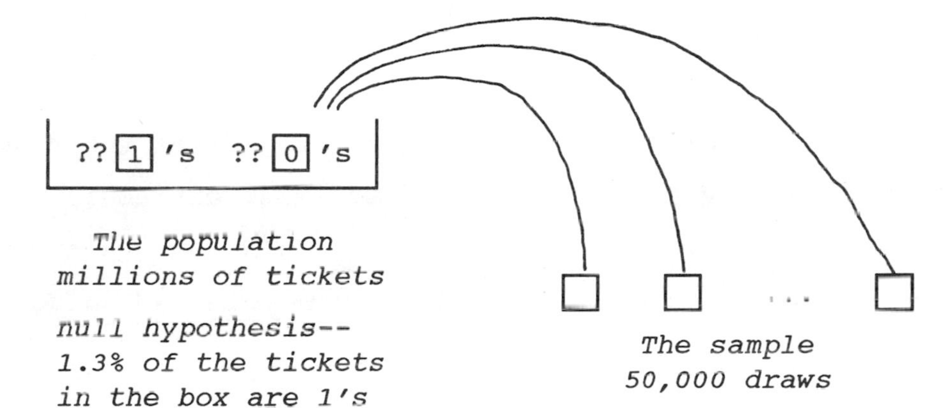

The percentage of 1's is expected to be 1.3%, and the SE is 0.05 of 1%. The percentage of 1's in the sample is 6.3%. The difference between 6.3% and 1.3% is almost impossible to explain as chance variation: $z \approx 100$, a record value. Apparently, many burglaries are not reported to the police.

11. (a) The total weight of the 1000 guineas in the Pyx will be like the sum of 1000 draws made at random with replacement from a box. The average of the numbers in the box is 128 grains, and the SD is $1/200 \times 128$ grains. (The numbers in the box represent the

11. (a) (continued)

possible weights of coins minted by the machine.) The expected value for the sum is 128,000 grains. The SE is $\sqrt{1000} \times 1/200 \times 128 \approx 20$ grains. The total weight of the coins in the Pyx will be 128,000 grains, give or take 20 grains or so, and it is almost impossible for the Master of the Mint to fail the Trial of the Pyx: the remedy of 640 grains is 32 SEs!

(b) The method is as in (a). The total weight of the coins in the Pyx will be 127,500 grains give or take 20 grains or so. Again, the Master of the Mint can hardly fail. The total weight is extremely unlikely to fall below 128,000 - 640 = 127,360 grains--a cutoff which is still 7 SEs below the expected value.

(c) 50,000 ± 200 grains.

Comment. The people who designed the Trial of the Pyx did not know about the square root law!

12. (a) There are 59 pairs, and in 52 of them, the treatment animal has a heavier cortex.

(b) On the null hypothesis, the expected number is $59 \times 0.5 = 29.5$ and the SE is $\sqrt{59} \times 0.5 \approx 3.84$. So 52 is nearly 6 SEs above average, and the chance is close to 0. Inference: treatment made the cortex weigh more.

(c) The average is 36 milligrams and the SD is 31 milligrams. The SE for the average is 4 milligrams, so $z = 36/4 = 9$ and $P \approx 0$. (This is like the tax example in section 1). Inference: treatment made the cortex weigh more.

12. (continued)

   (d) This blinds the person doing the dissection to the treatment status of the animal. It is a good idea, because it prevents bias; otherwise, the technician might skew the results to favor the research hypothesis.

Chapter 27. More Tests for Averages

1. No. The expected number of positives is 250, and the SE is $\sqrt{500} \times 0.5 \approx 11$. The observed number is 2.4 SEs above the expected. (Here, a one-sample test is appropriate.)

2. (a) The SE for the difference is 5.9%, so $z = 2.4/5.9 \approx 0.4$, looks like chance.

   (b) The SE for the difference is 1.8; the observed difference is 4.9; so $z = 4.9/1.8 \approx 2.7$ and $P \approx 0.3$ of 1%. The difference looks real.

Comment. The averages have more information.

3. Model: There are two boxes. The 1979 box has a ticket for each person in the population, marked 1 for those who had "a great deal or quite a lot" of confidence in the Supreme Court, and 0 for the others. The 1979 data are like 1000 draws from the 1979 box. The 1987 box is set up the same way. The null hypothesis says that the percentage of 1's in the 1987 box is the same as in the 1979 box. The alternative hypothesis says that the percentage of 1's has gone down.

3. (continued)

   The SD of the 1979 box is estimated from the data as $\sqrt{0.6\times0.4} \approx 0.49$. On this basis, the SE for the 1979 number is $\sqrt{1000}\times0.49 \approx 15$: the number of respondents in the sample who have "a great deal or quite a lot" of confidence in the Supreme Court is 600, but the chance error in that number is around 15. Convert to percent, relative to 1000. The SE for the 1979 percentage is estimated as 1.5%. The SE for the 1987 percentage is about 1.6%.

   The SE for the difference is computed from the square root law (p456) as $\sqrt{1.5^2+1.6^2} \approx 2.2\%$. The observed difference is 52 - 60 = -8%. On the null hypothesis, the expected difference is 0%. So $z = (\text{obs-exp})/\text{SE} = -8/2.2 \approx -3.65$, and $P \approx 1/10{,}000$. The difference looks real.

Comment. Either a one-sided or a two-sided test can be used; here, the distinction is not so relevant: it is discussed in chapter 29.

4. You can't tell. The method of section 2 does not apply, because you do not have two independent samples. The method of sections 3—4 does not apply, because you observe two responses for each person. See ex 4 on p468.

5. This is like the radiation-surgery example in section 4. Each subject has two possible responses, one to item A and one to item B; the investigators only observe one of the two, chosen at random. To make the test, pretend you have two independent random samples. With item A, the percentage who answer "yes" is 46%; the SE for this percentage is 3.5%. With item B, the percentage

5. (continued)

is 88% and the SE is 2.4%. The difference between the percentages is 46%−88% = −42%. The SE for the difference is conservatively estimated as $\sqrt{3.5^2+2.4^2} \approx 4.2\%$. So $z = -42/4.2 = 10$. The data say that framing the question makes a difference.

6. This is just like the previous exercise. In the calculator group, 7.2% get the right answer; in the pencil-and-paper group, 23.6%. The SEs are 1.6% and 2.7%. The difference between the percentages is −16.4%, and the SE for the difference is conservatively estimated as $\sqrt{1.6^2+2.7^2} \approx 3.1\%$. So $z = -16.4/3.1 \approx 5$, and $P \approx 0$. The difference is real. (Students who used the calculator seemed to forget what the arithmetic was all about.)

7. (a) No. This is like the radiation-surgery example in section 4. (Also see review exercises 5 and 6 above.) The SE for the treatment percent is 2.0%; for the control, 4.0%; for the difference, 4.5%. The observed difference, 0.9 of 1%, is only 0.2 of an SE. The difference is due to chance.

   (b) This is like example 4; the SE for the treatment average is 0.7 weeks; for the control average, 1.4 weeks; for the difference, 1.6 weeks. The observed difference is −7.5 weeks. So $z = -7.5/1.6 \approx 4.7$ and $P \approx 0$. The difference is real. Income support makes the released prisoners work less.

8. The data can be summarized as follows:

	Prediction Request	*Request only*
Predicts	22/46	NA
Agrees	14/46	2/46

(a) This is like the radiation-surgery example in section 4. The two percentages are 47.8% and 4.4%. The SEs are 7.4% and 3.0%. The difference is 43.4% and the SE for the difference is 8%. So $z = 43.4/8 = 5.4$ and $P \approx 0$. The difference is real. People overestimate their willingness to do volunteer work.

(b) The two percentages are 30.4% and 4.4%; the SEs are 6.8% and 3.0%. The difference is 26% and the SE for the difference is 7.4%. So $z = 26/7.4 = 3.5$ and $P \approx$ 2/10,000. The difference is real. Asking people to predict their behavior changes what they will do.

(c) Here, a two-sample z-test is not legitimate. There is only one sample, and two responses for each person in the sample. Both responses are observed, so the method of section 4 does not apply. The responses are dependent, so the method of example 3 does not apply. That is why the investigators compare 22/46 with 2/46.

Comment. In parts (a) and (b), the number of draws is small relative to the number of tickets in the box. So there is little difference between drawing with or without replacement, and little dependence between the treatment and control averages. See p463.

9. Parts (a–d) can be handled like example 4; parts (e–g), like the radiation-surgery example in section 4.

   (a) $z = 0.1/0.13 = 0.8$ and $P \approx 21\%$. At baseline, the difference between the two groups is well within the range of chance; the randomization worked. (Compare ex 7 on p20 and ex 3 on p465.)

   (b) $z = -3.1/0.15 \approx -21$ and $P \approx 0$. The intervention really got the blood pressure to go down.

   (c) $z = 0.3/0.65 \approx 0.45$ and $P \approx 33\%$. At baseline, the difference between the two groups is well within the range of chance; the randomization worked.

   (d) $z = -4.8/0.69 \approx -7$ and $P \approx 0$. The intervention really got the serum cholesterol to go down.

   (e) $z = 0.3/0.87 \approx 0.35$ and $P \approx 36\%$. At baseline, the difference between the two groups is well within the range of chance; the randomization worked.

   (f) $z = -13.3/0.86 \approx -15$ and $P \approx 0$. The intervention really got them to give up smoking.

   (g) The 6-year death rate in the treatment group was 3.28%, compared to 3.40% in control. The SEs are 0.22 of 1% and 0.23 of 1%. The difference is −0.12 of 1%, and the SE for the difference is 0.32 of 1%. So $z = -0.12/0.32 \approx -0.4$, and $P \approx 34\%$. The intervention got the risk factors down, but didn't change the death rate.

Comment. The sample sizes used in the calculation are at baseline; no adjustment is made for mortality. (This would be minor.)

10. This is like the radiation-surgery example in section 4. In the positive group, the percentage accepted is 28/53 × 100% ≈ 52.8%; in the negative group, 14.8%. The SEs are 6.9% and 4.8%. The difference is 38% and the SE for the difference is 8.4%. So z = 38/8.4 ≈ 4.5 and P ≈ 0%. There is a big difference between the two groups, and the difference cannot be explained by chance; journals prefer positive articles.

11. This test is not legitimate. The sample was not chosen at random; and there is dependence between the first-borns and second-borns.

Chapter 28. The Chi-Square Test

1.

Observed	*Expected*
1	7.6
10	30.1
16	7.4
35	6.9

$\chi^2 \approx 154$ on 3 degrees of freedom, so P ≈ 0 and option (ii) is right.

Comments. (i) Judges prefer well-educated grand jurors. (ii) The expecteds do not have to be whole numbers. For instance, if you roll a die 100 times, the expected number of aces is 16.666....

2. The expecteds are as follows:

	Married	*Widowed, divorced, or separated*	*Never married*
Employed	623	136	114
Unemployed	29	6	5
Not in labor force	48	10	9

(Entries are rounded, so the sum in column 2 is 152 not 153; the total in row 2 is a bit off as well.)

$\chi^2 \approx 20$ on 4 degrees of freedom, so $P \approx 5/1000$. This does not look like chance variation. The married men do better at getting jobs. (Or, men with jobs do better at getting married: the χ^2-test will not tell you which is the cause and which is the effect.)

3. The observed frequencies are too close to the expected ones for comfort: $\chi^2 \approx 2$ on 10 degrees of freedom, so $P \approx 0.4$ of 1% (left tail). This individual seems to have very good control over the dice. Maybe you should decline his invitation to play craps.

4. $\chi^2 \approx 0.2$ on 2 degrees of freedom, $P \approx 90\%$, a good fit.

5. Make a χ^2-test. We are interested in the chances, not just the average. $\chi^2 \approx 2.6$ on 5 degrees of freedom, $P \approx 25\%$, a good fit.

6. (a) Chance: it's a probability histogram.
   (b) The chance that $5 \le \chi^2 < 5.2$, where χ^2 is computed from 60 rolls of a fair die.
   (c) The chance that $5 \le \chi^2 < 5.2$ is bigger than the chance that $4.8 \le \chi^2 < 5$.

Comment. The exact probability distribution of χ^2, with 60 rolls, is quite irregular. As the number of rolls goes up, the histogram gets closer to the curve.

7. With 9 degrees of freedom, P will be bigger. Reason: that curve has more area to the right of 15.

Chapter 29. A Closer Look at Tests of Significance

1. Question (i) only (pp509–10).
2. Yes (p495).
3. False. You have to take the sample size into account too. For example, suppose the first investigator gets an average of 52, and the second one gets an average of 51. The first investigator gets $z = (52-50)/1 = 2$ and $P \approx 5\%$. The second investigator gets $z = (51-50)/0.33 = 3$ and $P \approx 0.3$ of 1%.
4. Yes: data snooping (p495).
5. It is hard to make sense out of "statistical significance" here, because there is no reasonable chance model for the data. The inner planets do not form a sample, they are the inner planets; similarly for the outer ones. (See ex 2 on p508.)
6. No. There seems to be a big effect, but one that is poorly estimated (p501; ex 7 on p502).
7. The concept of statistical significance does not apply, because the data are for the whole population, not a sample (p506). The difference is practically significant. The center of population is shifting to the West, and that makes a lot of difference to the economy and to the political balance of the country.

8. (a) The question makes sense; the data are from probability samples.

   (b) No. You need to know about the design of the samples, and their sizes (section 22.5).

   (c) Yes. Use the method of example 3 on p459. The SE for the 1975 sample percentage is 0.22 of 1%, and the SE for 1985 is about the same. The SE for the difference is 0.31 of 1%, so $z = 8.4/0.31 \approx 27$, and $P \approx 0$. This difference is off the chance scale, and very important in practical terms as well.

9. (a) The question makes sense, and the difference in attitudes is important. (This is a practical judgment, not a statistical one.)

   (b) The question makes sense, because the data are based on probability samples; to answer it, you need to know more about the design of the samples (section 22.5).

   (c) Now this is like example 3 on p459. The SE for the 1970 percentage is 1.6%; for 1985, 1.3%. The difference is −26%, and the SE for the difference is 2%, so $z = -26/2 = -13$ and $P \approx 0$.

10. These investigators seem to have made a number of statistical errors. For one thing, they made lots of tests. Even if all their null hypotheses were right, they were almost bound to find some highly significant differences. For another thing, they seem to be thinking

10. (continued)

that P measures the size of the effect, and it doesn't (p501). The effect they estimated is minute: a 100-fold increase in asbestos fiber concentration only increases the risk of lung cancer by a factor of 1.05. Finally, and most important: smoking is a major cause of lung cancer, and the investigators paid no attention to this variable.

The argument is weak, and there is no reason to think that asbestos in the water causes lung cancer.

Comment. The investigators found no effect for blacks or women; see note 38 to chapter 29.

11. $P \approx 5.9\%$ is pretty weak evidence; two-sided, $P \approx 11.8\%$, which is worse. Even to get these P-values, some data-snooping was needed. The argument is not good.

Comment. There were also serious problems with the model; see note 39 to chapter 29.

12. The statistical tests do not make much sense, because the data are for a whole population (p506).

Comment. Belmont and Marolla did a first-rate piece of work, but the tests are purely ceremonial.

Statistics 2 Freedman-Purves
Fall 1977

Diagnostic Quiz

[This test was given to 316 students.]

This is a diagnostic quiz to help us determine the general level of mathematical ability in the class. Many of the skills tested in the quiz will not be used in the course. Your score on this quiz will not affect your grade, but please do your best. If you do not know the answer to a question, do not guess--just leave it blank.

1. 300 is what percent of 2,000? [74% got this right.}

2. A town has 100,000 families; 0.1 of 1% of these families have incomes over $75,000 a year. The number of such families is ____________. [71% got this right.]

3. There are 100 million eligible voters in the United States. The Gallup poll interviews 5,000 of them. This amounts to one eligible voter out of every _________.
[61% got this right.]

4. In the United States, 1 person out of every 200 is in the army, and 8 of every 10,000 are army officers. What percentage of army personnel are officers, or can this be determined from the information given?
[20% got this right.]

5. In the United States, 1 person out of every 500 is in the navy, and one-sixth of naval personnel are officers. What fraction of the U.S. population consists of naval officers? Or can this be determined from the information given? [27% got this right]

6. $\sqrt{100,000}$ is about:

   (i) 30 (ii) 300 (iii) 1,000 (iv) 3,000 (v) can't tell

   [64% got this right.]

	True	False	Don't Know
7. $\sqrt{17}/17 = 17/\sqrt{17}$. [81% got this right.]	_____	_____	_____
8. $\sqrt{1/2}$ is smaller than 1/2. [41% got this right.]	_____	_____	_____

	True	False	Don't Know
9. $\sqrt{(2.5)^2 + (3.4)^2} = 2.5 + 3.4$	_____	_____	_____

[59% got this right.]

10. Solve for x and y, if possible: $x + 3y = 1$, $2x + y = -3$.

[54% got this right.]

11. A quart of vodka is 40% alcohol. Write a formula for the percentage of alcohol in a mixture of V quarts of vodka and J quarts of orange juice.

[15% got this right.]

12. This year John's mother is exactly three times as old as he is. Next year, their ages will add up to 50. How old is John? [64% got this right.]

13. Here is a quadratic equation: $3x^2 + 17x - 28.9207 = 0$ One of the following is a solution. Which one?

(i) 0.87 (ii) 1.37 (iii) 2.17 (iv) 3.81

[39% got this right.]

14. The graph of a straight line is shown below. The line has the equation $y = \frac{1}{3}x + 2$. Does the point (5.1, 3.6) lie on the line, or can this be determined from the information given? [49% got this right.]

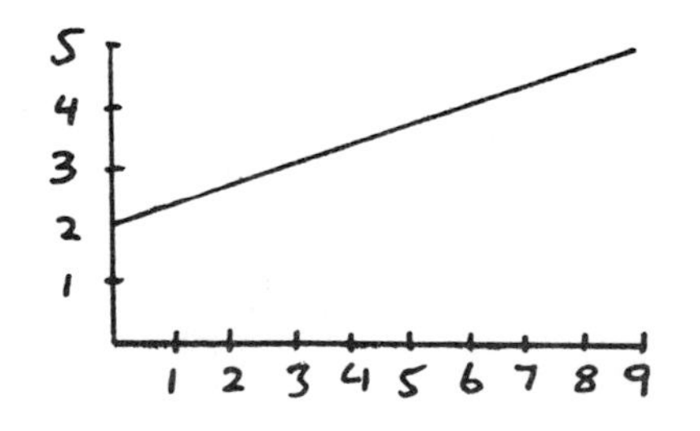

15. Someone is going to drive at a constant speed from San Francisco to Los Angeles by way of Palo Alto. The driver is wondering what speed to choose. Here are four factors of interest:

I	the distance from San Francisco after 1 hour
II	the time required to go 100 miles
III	the distance to Palo Alto after 1 hour
IV	the distance to LA after 1 hour

Distance is measured along the highway. Below are six graphs. Each point on a graph represents a whole trip, the constant speed of the car being shown along the horizontal axis. In four of the graphs, one of the factors listed above is plotted along the vertical axis.

Match the graph with the factor.

I	a	b	c	d	e	f
II	a	b	c	d	e	f
III	a	b	c	d	e	f
IV	a	b	c	d	e	f

[6% got this right.]

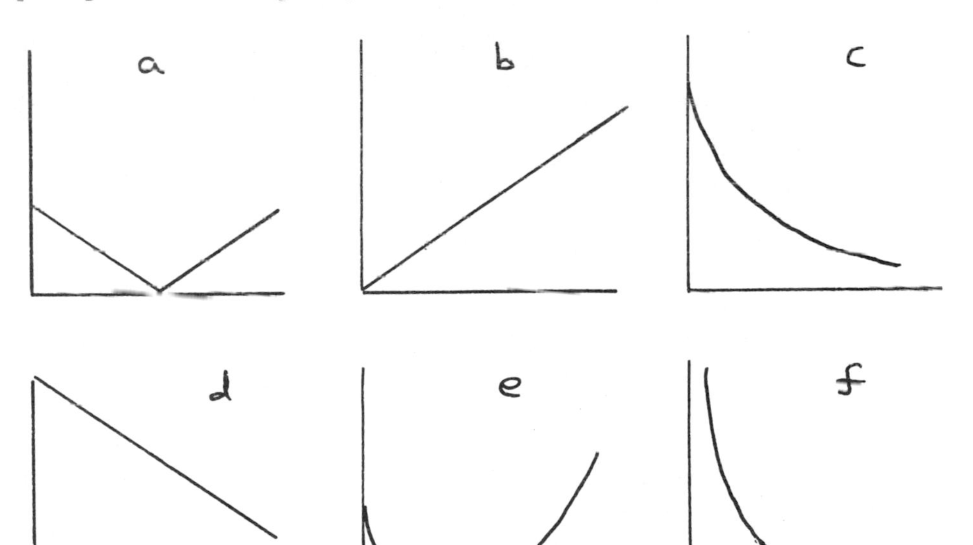

Probability

16. One of the two boxes below will be shaken, and a marble will be drawn out at random. If it is red you will win $1.

Box A		Box B	
	9 red marbles 1 blue marble		90 red marbles 10 blue marbles

(i) Box A is better. [8% chose this option.]

(ii) Box B is better. [7% chose this option.]

(iii) Both boxes offer the same chance of winning. [80% chose this option.]

17. You throw one die. What is the chance of getting an ace ⚀? [83% got this right.]

18. You throw a pair of dice. What is the chance of getting two aces ⚀⚀? [48% got this right.]

19. A coin will be tossed either 2 times or 100 times. You will win $2 if the number of heads is equal to the number of tails, no more and no less.

(i) 2 tosses is better. [28% chose this option.]

(ii) 100 tosses is better. [21% chose this option.]

(iii) Both offer the same chance of winning. [45% chose this option.]

20. Here are two situations:

A) A coin will be tossed 100 times. If it comes up heads 60 or more times you win $1.

B) A coin will be tossed 1,000 times. If it comes up heads 600 or more times you will win $1.

(i) Situation A is better. [21% chose this option]

(ii) Situation B is better. [9% chose this option.]

(iii) Both offer the same chance of winning. [64% chose this option.]

[The percents in 16, 19, 20 do not add to 100%, because about 5% of the students did not choose any option.]

Calculus

Have you had a college calculus course? Yes 37% No 63%

If yes, please work this section.

21. Differentiate x^3. [83% got this right.]

22. Find $\int_{-1}^{2} x^3 \, dx$. [28% got this right.]

23. Solve $\frac{dy}{dx} = x$. [21% got this right.]

24. To find the c which minimizes $(x_1 - c)^2 + (x_2 - c)^2 + (x_3 - c)^2$:

    (i) Differentiate with respect to x.
    (ii) Differentiate with respect to c.
    (iii) Can't do it by calculus.

    [22% chose option (i), 19% chose option (ii), 15% chose option (iii), and 44% declined to choose.]

[In this section, percents are based on the 117 students who said they had a college calculus course.]

Note: the average score on questions 1 through 20 was 10, with an SD of 4. In statistics 20 (which has a calculus prerequisite and is aimed at students in quantitative fields), the average score on similar tests was around 13 out of 20, with the same SD of 4.

Statistics 2 Freedman-Purves
Fall 1977

Midterm

[This test was given to 311 students.]

PRINT YOUR NAME ______________________________

SIGN YOUR NAME ______________________________

LECTURE TIME: 12-1 OR 1-2?

TA'S NAME ______________ LAB TIME ____________

To get full credit, you must show work.

x	0.1	0.2	0.3	0.4	0.5	0.6	0.7	0.8	0.9
$\sqrt{x}$	0.32	0.45	0.55	0.63	0.71	0.77	0.84	0.89	0.95

z	0.1	0.2	0.25	0.50	0.75	1.00	1.25	1.60	1.75	2.00	2.25
A(z)	8%	16%	20%	38%	55%	68%	79%	90%	92%	95%	98%

1. In a large lecture course, the scores on the final examination followed the normal curve closely. The average score was 60 points and three-fourths of the class scored between 50 and 70 points. The SD of the scores was

   (i) larger than 10 points
   (ii) smaller than 10 points
   (iii) impossible to say with information given

   Explain your answer. (10 points)
   [82% got this right.]

2. The figure below is a histogram for the blood pressures of subjects in a certain study. The percentage who had blood pressures between 120 mm and 140 mm is exactly equal to the area under __________ between those two values. Fill in the blank with one of the two options: histogram, normal curve. (10 points)

[68% got this right.]

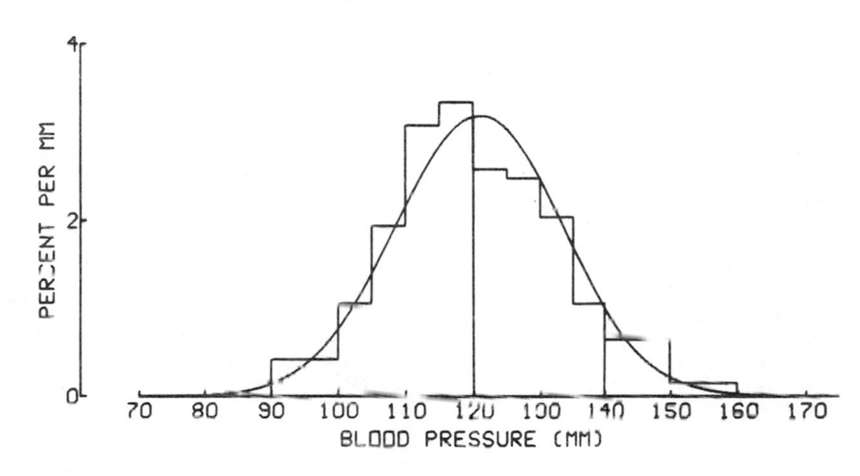

3. Find the correlation coefficient for the data set below. (20 points)

[78% got this right.]

x	y
4	4
4	2
3	1
3	5
1	5
3	7

4. Pearson and Lee obtained the following results for about 1,000 men:

average height ≈ 69 inches, SD ≈ 2.5 inches

average forearm length ≈ 18 inches, SD ≈ 1 inch, r ≈ 0.8

Of the men who were 6 feet tall (to the nearest inch), about what percentage had forearms shorter than 18 inches? (20 points).

[49% got this right.]

5. In a study of a representative group of men, the correlation between height and weight was 0.43. One man in the study was both one SD above average in height and one SD above average in weight. His weight will be

   (i) larger than (ii) smaller than (iii) equal to

   the average weight of all men of his height in the study. Explain your choice. (20 points)

   [61% got this right.]

6. The figure below is a histogram for the scores on the final in a certain class. Find the 75th percentile of these scores. (20 points)

   [61% got this right.]

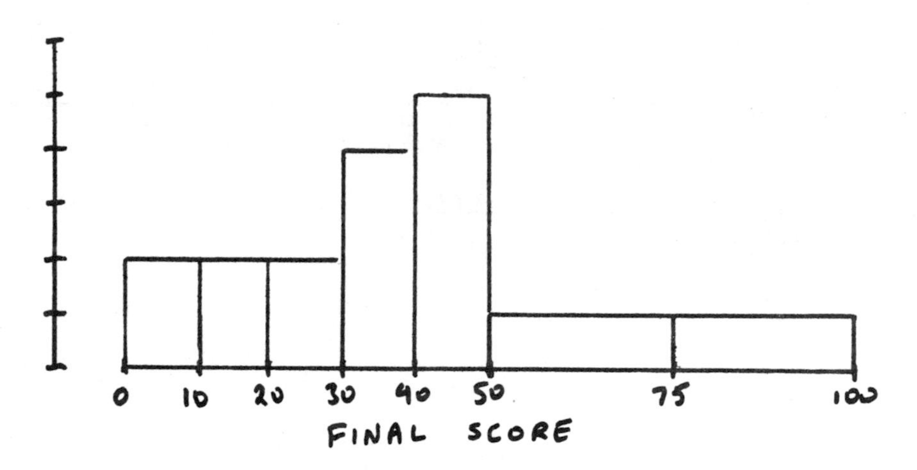

Psychology 60

Final Examination

December 12, 1977

[This course was taught by Professor J. Merrill Carlsmith, Stanford University, using a trial edition of our book. The final is reproduced with his permission. Some questions test material covered in lectures but not in the text. The exam was taken by about 150 students: the median score was 78 out of 100.]

Each question is worth 10 points.

1. 60% of the pupils in a particular university are male. What is the probability that a random sample of 400 pupils will contain fewer than 220 males?

2. A random sample of 6 Stanford faculty members gave the following data for number of publications and salary.

Faculty member	Number of Publications	Salary (in 000's)
A	16	20
B	16	25
C	20	22
D	20	30
E	24	28
F	24	25

i) Write the equation of the regression line.

ii) If we found another faculty member whose number of publications was 22, what would we guess his salary to be?

iii) If in a burst of excitement upon seeing these data, I quickly published 5 more papers, how much would you expect my salary to go up?

iv) How would you interpret the intercept of the regression line?

3. I propose to you a new game. You roll 2 dice. If the sum of the numbers showing is either 6, or 7, or 8, I win. If it is 2, 3, 4, 5, 9, 10, 11, 12, you win. Since you have lots more possible winning combinations than I do, the rules are that you pay me $2.00 when I win and I pay you $1.00 when you win. If we play this game 30 times, how much do you think you will win or lose? (I will be in my office this afternoon for anyone who feels like playing.)

4. In an experiment to test the efficacy of Vitamin C in preventing colds, 9 experimental subjects take Vitamin C every day for 1 year, while 4 control subjects take no Vitamin C. The number of colds are tabulated at the end of the year. The thing we are interested in is the size of the reduction in the number of colds which can be attributed to Vitamin c.

   i) Give an interval which we can be 95% sure covers the true reduction.

   ii) Does Tukey's quick and dirty test suggest that Vitamin C had an effect?

Experimentals		Controls
0	3	2
0	0	6
1	1	2
2	1	2
1		

5. A certain town has 25,000 families. The average number of children per family is 2.6, with an SD of 0.80. The distribution is not normal, however, since 25% of the families have no children at all. If we draw a random sample of 90 families, what are the chances that between 23% and 27% of the sample families will have no children?

6. Last year baseball instituted what was called the "free agent draft." In essence, this gave players the right to negotiate a contract with any team they chose to, rather than belonging to a particular team forever or until traded. A few of the wealthiest clubs (like the Yankees) promptly paid enormous sums to obtain a few players who had had very high batting averages the previous year. During the course the year, sports writers had lots of fun pointing out what bad judgement the owners had shown, for almost none of these high-priced players did as well this year as they had done the year before. Do you think all that money made the players fat and lazy? Or do you have another explanation?

7. A large number of measurements on a standard kilogram have established that our weighing procedure gives an average which is 500 micrograms too high, with an SD of 10 micrograms. We have just been sent a new checkweight which we have been asked to weigh. The owners of this checkweight specify that they wish the weight we report to be accurate to within <u>1</u> micrograms. We reply that we can't guarantee that, but that we are prepared to guarantee that our answer will be accurate to within 1 micrograms 95% of the time. How many measurements do we need to take?

8. It is known that nationally, 10% of all lawyers are female. A random sample of lawyers in a particular state yielded 400 males and 100 females. 80% of the sampled male lawyers favored passage of the Equal Rights Amendment, while 90% of the sampled female lawyers favored its passage. Is the difference between male and female lawyers in this state real, or is it just chance variation?

9. First-born children are less likely to become alcoholic than are later-born children. I wonder whether this fact is also true of birth order with pairs of twins. To study this, I find 8 pairs of twins, and classify each twin as either first-born or second-born. The measure I use is the average daily ingestion of alcohol. Formulate an appropriate null hypothesis and do a test of significance.

		ALCOHOL INGESTED (in ounces)	
		1st born	2nd born
Twin Pair Numbers	1	4	5
	2	0	3
	3	2	2
	4	0	1
	5	5	4
	6	3	5
	7	4	5
	8	1	2

10. A study of 500 babies looked at the relationship between their weight at birth and the age at which they first slept through the night. The birth weights averaged out to 90 oz. with an SD of 15 oz. The ages at which they first slept all night averaged 50 days with an SD of 10 days. The correlation between the two variables was -.60. If we draw a sample of 16 babies who weighed 105 oz. at birth, what is the probability that the average age at which these 16 babies sleep through the night is less than 40? (Hint: You have never seen a problem like this one before, although you have seen all of the component parts.)

** (For thinking about over the holidays only). You are taking a plane trip and have heard that the odds against someone bringing a bomb on board are 1000 to 1. You are little worried, but then you read that the odds against <u>two</u> people (independently) bringing a bomb on the same plane are 1,000,000 to 1. What should you do? (Bring a bomb on board?)

Statistics 2 | Freedman-Purves
Fall 1977

FINAL

[This three-hour test was taken by 294 students.]

Print your name ______________________________

Sign your name ______________________________

Lab time ______________________________

Your TA's name ______________________________

Your instructor's name ______________________________

To get full credit, you must show work. No work, no credit. There is only one exception: problem #5.

square root table

x	$\sqrt{x}$
0.10	0.32
0.15	0.39
0.20	0.45
0.22	0.47
0.25	0.50
0.30	0.55
0.35	0.59
0.40	0.63
2.70	1.64
2.80	1.67
2.90	1.70
18.00	4.24
25.00	5.00

normal table

z	A(z)
0.05	4%
0.075	6%
0.10	8%
0.125	10%
0.20	16%
0.25	20%
0.50	38%
0.75	55%
1.00	68%
1.25	79%
1.50	87%
2.00	95%
2.50	99%

1. Commute Distance (4 points)

As part of a survey, one large manufacturing company asked a thousand of its employees how far they had to commute to work each day (round trip). The data was analyzed by computer and on the printout the average round trip commute distance was reported as 11.3 miles, with an SD of 16.2 miles. Would a rough sketch of the histogram for the data look like (i) or (ii) or (iii)? Or is there a mistake somewhere? Explain your answer. [55% got this right.]

(i) (ii) (iii)

2. Children's Heights (4 points)

A large sample of children was followed over time. One investigator looked at all the children who were at the 90th percentile in height at age four. Some of these children turned out to be above the 90th percentile in height at age eighteen, and others were below. The number who were above was

(i) quite a bit smaller than
(ii) about the same as
(iii) quite a bit larger than

the number who were below. Or is more information needed? Give a reason to support your answer. (You may assume that the scatter diagram is football-shaped.)
[46% got this right.]

3. The Box (8 points)

A hundred draws are made at random with replacement from the box | 0 0 0 1 2 |. Estimate the chance that 1 turns up on exactly 20 draws. Show your work. [57% got this right.]

4. Survey Research (4 points)

A survey research center conducts frequent opinion polls, using large samples drawn by probability methods which are practically free from bias. Each of the last 100 polls was carried out to estimate a percentage. All the standard errors were computed by the appropriate technique, and turned out to be very close to 3 percentage points. About how many of the estimates were off by more than 3 percentage points? Explain briefly. [67% got this right.]

5. Blanks (4 points)

Fill in the blanks, using one word from each pair below, to make up two true sentences. Write both sentence down.

"If two things are __(i)__, and you want to find the chance that __(ii)__ will happen, you can __(iii)__ the chances."

(i) incompatible, independent
(ii) both, at least one
(iii) add, multiply

(In this problem, no work need be shown.)

[77% got this right.]

6. The Speed of Light (8 points)

The speed of light is measured 25 times by a new procedure. The 25 measurements are recorded, and show no trend or pattern. Then the investigators work out the average and SD of the 25 numbers; the average is 299,789.2 kilometers per second and the SD is 12 kilometers per second.

(a) Find an approximate 95% confidence interval for the speed of light, showing your work. (You may assume the Gauss model, with no bias.)

(b) Now the investigators measure the speed of light a 26th time by the same procedure, and get 299,781 kilometers per second. Is this a surprising result?

Yes _____ No _____

Check one, and give your reason.

[65% got this right.]

7. Southern California (4 points)

Los Angeles has about four times as many registered voters as San Diego. A simple random sample of registered voters is taken in each city, to estimate the percentage who will vote for school bonds. Other things being equal, a sample of 4,000 taken in Los Angeles will be about

(i) four times as accurate
(ii) twice as accurate
(iii) as accurate

as a sample of 1,000 taken in San Diego. Choose one option and say why. [39% got this right.]

8. Hospitals (4 points)

One hospital has 218 live births during the month of January. Another has 536. Which is likelier to have 55% or more male births? Or is it equally likely? Explain. (There is about a 52% chance for a live-born infant to be male.)
[67% got this right.]

9. The Surveyor (8 points)

A surveyor is measuring the distance between five points A, B, C, D, E. They are all on a straight line. He finds that each of the four distances AB, BC, CD, and DE measures one mile, give or take an inch or so. These four measurements are made independently, by the same procedure.

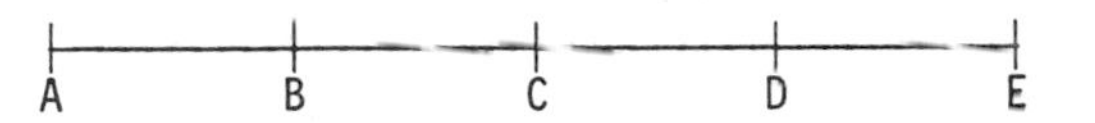

The distance from A to E is about four miles, give or take around

4 inches   2 inches   1 inch   1/2 inch   1/4 inch

Explain briefly. (You may assume the Gauss model, with no bias.)

[56% got this right.]

10. Television (8 points)

In a certain town, there are 25,000 households. On the average, there are 1.2 color TV sets per household, with an SD of 0.6; but 10% of the households do not have color TV. As part of the a market survey, a simple random sample of 900 households is drawn from the 25,000. What is the chance (approximately) that somewhere between 9% and 11% of the sample households will not have color TV? Show your work. [64% got this right.]

11. Carnegie (12 points)

There are about 2,700 institutions of higher learning in the United States (including junior colleges and community colleges). In 1976, as part of a continuing study of higher education, the Carnegie Commission took a simple random sample of these institutions. The average enrollment in the 225 sample schools was 3,700, with an SD of 6,000. A histogram for the enrollments was plotted and did not follow the normal curve. However, the average enrollment at all 2,700 institutions was estimated to be around 3,700, give or take 400 or so. Say whether each of the following statements is true or false, and explain why. [62% got this right.]

(a) It is estimated that 95% of the institutions of higher learning in the United States enroll between 3,700 - 800 = 2,900 and 3,700 + 800 = 4,500 students.

(b) An approximate 95%-confidence interval for the average enrollment of all 2,700 institutions runs from 2,900 to 4,500.

(c) If someone takes a simple random sample of 225 institutions of higher learning, and goes two SEs either way from the average enrollment of the 225 sample schools, there is about a 95% chance that this interval will cover the average enrollment of all 2,700 schools.

(d) The normal curve can't be used to figure confidence levels here at all, because the data doesn't follow the normal curve.

12. High Schools (8 points)

There are about 25,000 high schools in the United States; each high school has a principal. As part of a national survey of education, a simple random sample of 225 high schools is chosen. In 202 of the sample high schools the principal has an advanced degree.

(a) If possible, find an approximate 95% confidence interval for the percentage of all 25,000 high school principals who have advanced degrees, showing your work. If this is impossible, explain why.

The 25,000 high schools in the United States employ a total of about one million teachers. As it turned out, the 225 sample high schools employed a total of 10,000 teachers, of whom 5,010 had advanced degrees.

(b) If possible, find and approximate 95% confidence interval for the percentage of all one million high school teachers with advanced degrees, showing your work. If this is impossible, explain why.

[62% got this right.]

13. Reading (8 points)

The National Assessment of Educational Progress (NAEP) administered a reading test to a nationwide probability sample of 9-year-olds in 1971. The same test was administered to an independently chosen sample of 9-year-olds in 1975. There appears to have been some improvement: in 1971, the average score was 67.2 out of 100, while in 1975 the average score was 68.5 out of 100. Or can this be explained as a chance variation? Explain your reasoning.

You may assume the NAEP took independent simple random samples in 1971 and 1975; there were 1,600 children in each sample, and in both years the SD of the scores was very nearly 14 points out of 100.

[53% got this right.]

14. Belmont and Marolla (8 points)

Belmont and Marolla conducted a study on the relationship between birth order, family size, and intelligence. The subjects consisted of all Dutch men who reached the age of 19 between 1963 and 1966. These men were required by law to take the Dutch army induction tests, including Raven's intelligence test. The results showed that for each family size, measured intelligence decreased with birth order: first-borns did better than second-borns, second-borns did better than third-borns, and so on. And for any particular birth order, intelligence decreased with family size: for instance, first-borns in two-child families did better than first-borns in three-child families. These results remained true even after controlling for the social class of the parents. Taking, for instance, men from two-child families:

- the first-borns averaged 2.575 on the test;
- the second-borns averaged 2.678 on the test.

(Raven test scores range from 1 to 6, with 1 being best and 6 worst.) The difference is small, but if it is real, it has interesting implications for genetic theory. To show that the difference was real, Belmont and Marolla made a two-sample z-test. The SD for the test scores was around one point, both for the first-borns and the second-borns, and there were 30,000 of each, so

$$\text{SE for sum} \approx \sqrt{30{,}000} \times 1 \text{ point} \approx 173 \text{ points}$$

$$\text{SE for average} \approx 173/30{,}000 \approx 0.006 \text{ points}$$

$$\text{SE for difference} \approx \sqrt{(0.006)^2 + (0.006)^2} \approx 0.008 \text{ points.}$$

Therefore, $z = (2.575 - 2.678)/0.008 \approx -12.6$, and P is astonishingly small. Belmont and Marolla concluded:

> Thus the observed difference was highly significant... a high level of statistical confidence can be placed in each average because of the large number of cases.

Was it appropriate to make a two-sample z-test in this situation?

Yes _____ No _____

Check one, and justify it.

[34% got this right.]

15. The Effects of Exercise (8 points)

An investigator in the Statistics Department of a large university is interested in the effect of exercise in maintaining mental ability. He decides to study the faculty members aged 40 to 50 at his university, looking separately at two groups: The ones that exercise regularly, and the ones that don't. There turn out to be several hundred people in each group, so he takes simple random sample of 25 persons from each group, for detailed study. One of the things he does is to administer an IQ test to the sample people, with the following results:

	regular exercise	no regular exercise
sample size	25	25
average score	135	121
SD of scores	15	15

The difference between the averages is "highly statistically significant." The investigator concludes that exercise does indeed help to maintain mental ability among the faculty members aged 40 to 50 at his university. Is this conclusion justified?

Yes _____ No _____

Check one, and say why.

[59% got this right.]

DIAGNOSTIC QUIZ — MR. FREEDMAN
STATISTICS 2/20 — FALL 1988 — MR. PURVES

This quiz was taken by 23 students in Statistics 20 and 251 students in Statistics 2. A very similar quiz was taken by 316 students in Statistics 2, Fall 1977: see pp147-51. (Statistics 20 has a calculus prerequisite and is taught in small sections.) Results are tabulated for all three courses: the "pass rate" is the percentage of students giving the right answer.

This is a diagnostic quiz to help us determine the general level of mathematical ability in the class. Many of the skills tested in the quiz will not be used in the course. Your score on this quiz will not affect your grade, but please do your best. If you do not know the answer to a question, do not guess--just leave it blank.

1. 300 is what percent of 2,000?

	Pass Rate
Statistics 20, Fall 1988	*87%*
Statistics 2, Fall 1977	*74%*
Statistics 2, Fall 1988	*74%*

2. A town has 100,000 families; 0.1 of 1% of these families have incomes over $75,000 a year. The number of such families is _____.

	Pass Rate
Statistics 20, Fall 1988	*83%*
Statistics 2, Fall 1977	*71%*
Statistics 2, Fall 1988	*63%*

3. There are 100 million eligible voters in the United States. The Gallup poll interviews 5,000 of them. This amounts to 1 eligible voter out of every _____.

	Pass Rate
Statistics 20, Fall 1988	*65%*
Statistics 2, Fall 1977	*61%*
Statistics 2, Fall 1988	*51%*

4. In the United States, 1 person out of every 200 is in the army, and 8 out of every 10,000 are army officers. What percentage of army personnel are officers, or can this be determined from the information given?

	Pass Rate
Statistics 20, Fall 1988	*30%*
Statistics 2, Fall 1977	*20%*
Statistics 2, Fall 1988	*15%*

5. In the United States, 1 person out of every 500 is in the navy, and one-sixth of naval personnel are officers. What fraction of the US population consists of naval officers? Or can this be determined from the information given?

	Pass Rate
Statistics 20, Fall 1988	*52%*
Statistics 2, Fall 1977	*27%*
Statistics 2, Fall 1988	*27%*

6. $\sqrt{100{,}000}$ is about:

(a) 30 (b) 300 (c) 1,000 (d) 3,000 (e) Can't tell

	Pass Rate
Statistics 20, Fall 1988	*74%*
Statistics 2, Fall 1977	*64%*
Statistics 2, Fall 1988	*57%*

7. $\sqrt{17}/17 = 17/\sqrt{17}$ (a) True (b) False (c) Don't know

	Pass Rate
Statistics 20, Fall 1988	*96%*
Statistics 2, Fall 1977	*81%*
Statistics 2, Fall 1988	*82%*

8. $\sqrt{.5}$ is smaller than .5 (a) True (b) False (c) Don't know

	Pass Rate
Statistics 20, Fall 1988	*44%*
Statistics 2, Fall 1977	*41%*
Statistics 2, Fall 1988	*36%*

9. $\sqrt{2.5^2+3.4^2} = 2.5+3.4$ (a) True (b) False (c) Don't know

	Pass Rate
Statistics 20, Fall 1988	*74%*
Statistics 2, Fall 1977	*59%*
Statistics 2, Fall 1988	*47%*

10. A quart of vodka is 40% alcohol. Write a formula for the percentage of alcohol in a mixture of V quarts of vodka and J quart of orange juice.

	Pass Rate
Statistics 20, Fall 1988	*48%*
Statistics 2, Fall 1977	*15%*
Statistics 2, Fall 1988	*8%*

11. This year John's mother is exactly three times as old as he is. Next year, their ages will add up to 50. How old is John?

	Pass Rate
Statistics 20, Fall 1988	*87%*
Statistics 2, Fall 1977	*64%*
Statistics 2, Fall 1988	*69%*

12. Here is a quadratic equation: $3x^2+17x-28.9207 = 0$. One of the following is a solution; which one?

    (a) 0.87 (b) 1.37 (c) 2.17 (d) 3.81

	Pass Rate
Statistics 20, Fall 1988	*74%*
Statistics 2, Fall 1977	*39%*
Statistics 2, Fall 1988	*39%*

13. The graph of a straight line is shown below. The line has the equation $y=.33x+2$. Does the point (5.1,3.6) lie on the line? Or can this be determined from the information given?

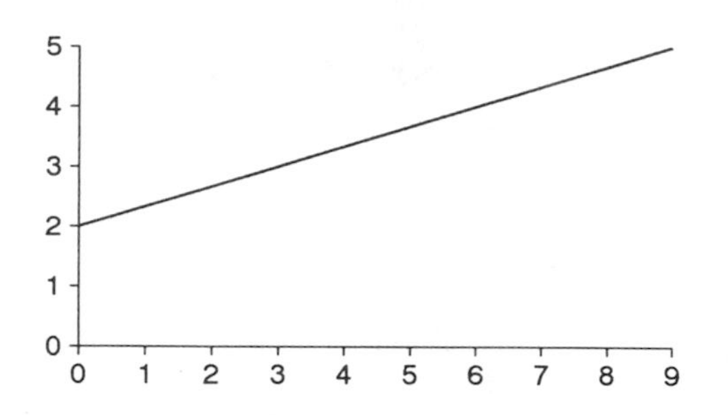

	Pass Rate
Statistics 20, Fall 1988	*74%*
Statistics 2, Fall 1977	*49%*
Statistics 2, Fall 1988	*61%*

14. You throw one die. What is the chance of getting an ace ⚀?

	Pass Rate
Statistics 20, Fall 1988	*96%*
Statistics 2, Fall 1977	*83%*
Statistics 2, Fall 1988	*88%*

15. You throw a pair of dice. What is the chance of getting two aces ⚀⚀?

	Pass Rate
Statistics 20, Fall 1988	*78%*
Statistics 2, Fall 1977	*48%*
Statistics 2, Fall 1988	*41%*

16. Here are two situations:

(i) A coin will be tossed 100 times. If it comes up heads 60 or more times, you win $1.

(ii) A coin will be tossed 1,000 times. If it comes up heads 600 or more times, you win $1.

Which is better? Or do they offer the same chance of winning?

	Pass Rate
Statistics 20, Fall 1988	*22%*
Statistics 2, Fall 1977	*21%*
Statistics 2, Fall 1988	*14%*

The denominators for pass rates in question 17-20 include all students in the class, whether or not they have had a college calculus course.

17. Differentiate x^2.

	Pass Rate
Statistics 20, Fall 1988	*83%*
Statistics 2, Fall 1988	*29%*

18. Find $\int_{-1}^{2} x^3 \, dx$.

	Pass Rate
Statistics 20, Fall 1988	*52%*
Statistics 2, Fall 1988	*14%*

19. Solve $\frac{dy}{dx} = x$.

	Pass Rate
Statistics 20, Fall 1988	*43%*
Statistics 2, Fall 1988	*10%*

20. To find the c which minimizes $(x_1-c)^2+(x_2-c)^2+(x_3-c)^2$:

(a) Differentiate with respect to x.
(b) Differentiate with respect to c.
(c) Can't do it by calculus.

	Pass Rate
Statistics 20, Fall 1988	*65%*
Statistics 2, Fall 1988	*9%*

MIDTERM
STATISTICS 20

MR. FREEDMAN
FALL 1989

[37 students took this test]

1. According to an observational study done at Kaiser Permanente in Walnut Creek, California, users of oral contraceptives have a higher rate of cervical cancer than non-users, even after adjusting for age, education, marital status, religion, and smoking. Investigators concluded that the pill causes cervical cancer. Were they right to do so? Answer yes or no, and explain briefly. *[84% got this right]*

2. The sketches show results of two studies on the pill, for women age 25-29. In one study, the pill adds about 10 mm to blood pressures; in the other, the pill adds about 10%. Which is which, and why? *[95% got this right]*

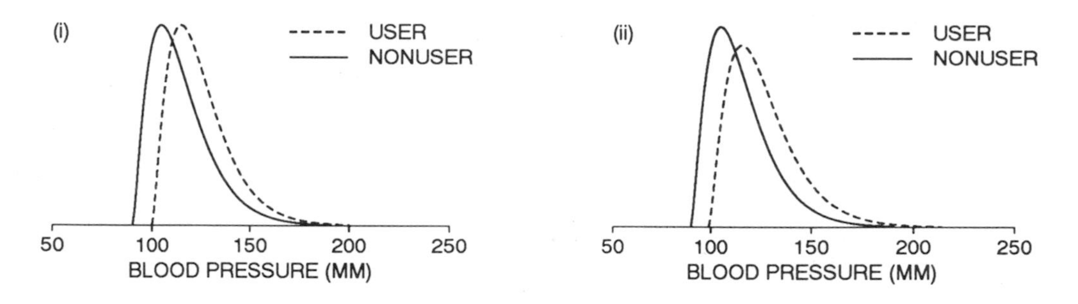

3. In 1983, investigators administered questionnaires to Russian emigres in New York. Subjects who described KGB leaders as "competent" also said they participated less often in political protests in Russia. The correlation remained after adjusting for age, education, and status. The investigators concluded:

   > ...persons who perceived the KGB to be highly competent were less likely to engage in unorthodox behavior... perceptions of the KGB's competence serve to deter would-be nonconformists.

   Is the conclusion justified? Answer yes or no, and explain briefly. *[32% got this right]*

4. The great French kings of history had mediocre chief ministers, while the great ministers served under kings of lesser talent. Is this a fact of French history? or of statistics? Discuss briefly. *[43% got this right]*

5. The figure below is a scatter plot of income against education, for a representative sample of men age 25–29 in Texas in 1988. Or is something wrong? Explain briefly. *[46% got this right]*

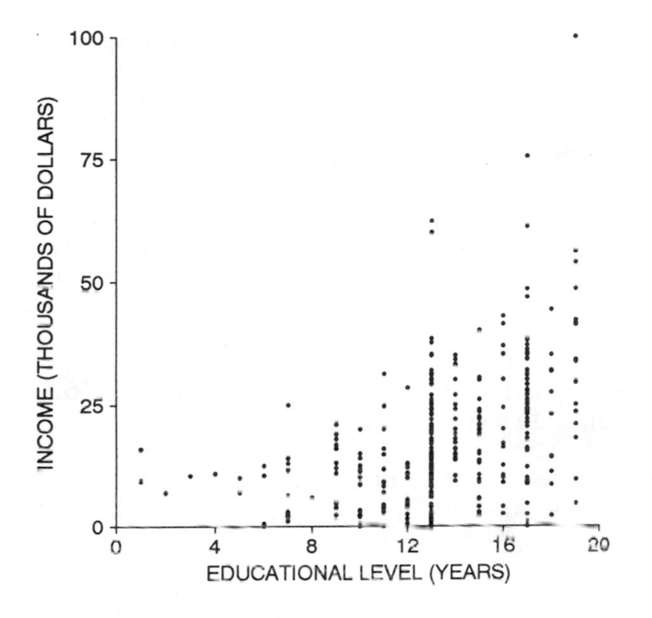

6. For women age 25–29 in California, the relationship between income and education can be summarized as follows.

   average education = 12 years, SD = 3.5 years
   average income = $11,600, SD = $10,500, r = 0.4

   (The scatter diagram has the same general shape as the one in problem 5 above).

   (a) Find the r.m.s. error of the regression line for predicting income from education.

   (b) Predict the income of a woman with 14 years of education.

   (c) This prediction is likely to be off by $________ or so. Or can this be determined from the information given?

   (d) Repeat parts (b) and (c), for a woman with 10 years of education.

   [89% got at least half credit on this question]

7. A coin will be tossed 10 times. Find the chance that there will be 2 heads among the first 5 tosses, and 4 heads among the last 5 tosses. *[73% got this right]*

FINAL
STATISTICS 20

MR. FREEDMAN
FALL 1989

[35 students took this three-hour test]

1. DES was given to pregnant women to prevent miscarriage. A literature review found 3 randomized controlled experiments, and 5 nonrandomized studies with control groups. The rate of miscarriages was about the same in the treatment groups, for all 8 studies. The rate was also about the same in the 3 control groups, for the randomized controlled experiments. However, the rate was substantially higher among the control groups in the 5 nonrandomized studies. How do you interpret these data? *[71% got this right]*

2. By Census definitions, a "family" consists of 2 or more related persons living together; a "household" has 1 or more people living in the same housing unit. In 1987, the average income for households was about 10% less than the average income for families. How can this be? Discuss briefly. *[86% got this right]*

3. A statistical analysis is made of the midterm and final scores in a large course, with the following results:

   average midterm score ≈ 60, SD ≈ 15
   average final score ≈ 65, SD ≈ 20, r ≈ 0.50

   The scatter diagram is football-shaped.

   (a) About what percentage of students scored over 80 on the final?

   (b) Of the students who scored 80 on the midterm, about what percentage scored over 80 on the final?

   [91% got at least half credit for this question]

4. The unconditional probability of event A is 1/3; the unconditional probability of B is 1/10. True or false, and explain:

   (a) If A and B are independent, they must also be mutually exclusive.

   (b) If A and B are mutually exclusive, they cannot be independent.

   [69% got at least half credit for this question]

5. A gambler plays roulette 100 times. There are two possibilities:

   (A) Betting $1 on a section each time.

   (B) Betting $1 on red each time.

   A section bet pays 2 to 1, and there are 12 chances in 38 to win. Red pays even money, and there are 18 chances in 38 to win. True or false, and explain:

   (a) The chance of coming out ahead is the same with A and B.

   (b) The chance of winning more than $10 is bigger with A.

   (c) The chance of losing more than $10 is bigger with A.

   [00% got at least half credit for this question]

6. Shown below are probability histograms for the sum of 100, 400 and 900 draws from the box

   | 99 $\boxed{0}$'s $\boxed{1}$ |.

   Which histogram is which? Why?

   [97% got this right]

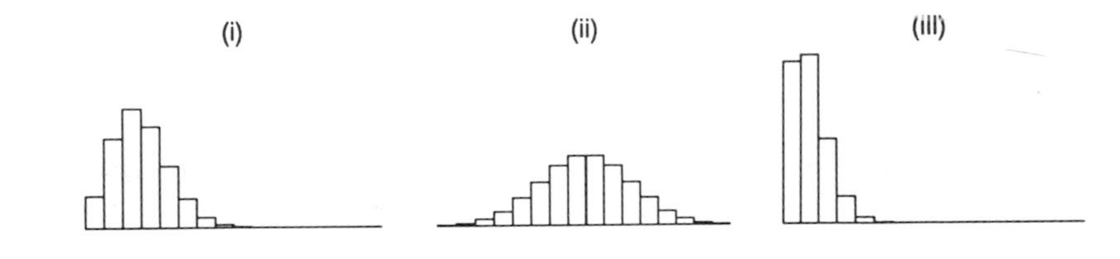

7. On Sunday, September 11, 1988, the *San Francisco Examiner* ran a story headlined--

   3 IN 10 BIOLOGY TEACHERS BACK BIBLICAL CREATIONISM

   Arlington, Texas. Thirty percent of high school biology teachers polled believe in the biblical creation and 19 percent incorrectly think that humans and dinosaurs lived at the same time, according to a nationwide survey published Saturday.

   "We're doing something very, very, very wrong in biology education," said Dana Dunn, one of two sociologists at the University of Texas, Arlington.

   Dunn and Raymond Eve sent questionnaires to 20,000 high school biology teachers selected at random from a list provided by the National Science Teachers Association and received 200 responses....

   The sociologists are doing something very, very, very wrong in statistics. What is that? *[97% got this right]*

8. One hospital has 218 live births during the month of January. Another has 536. Which is likelier to have 55% or more male births? Or is it equally likely? Explain. (There is about a 52% chance for a live-born infant to be male.) *[91% got this right]*

9. A survey organization takes a simple random sample of 625 households from a city of 80,000 households. On the average, there are 2.30 persons per sample household, and the SD is 1.75. Say whether each of the following statements is true or false, and explain.

   (a) The 2.30 is 0.07 or so off the average number of persons per household in the whole city.

   (b) A 95%-confidence interval for the average household size in the sample is 2.16 to 2.44.

   (c) A 95%-confidence interval for the average household size in the city is 2.16 to 2.44.

   (d) 95% of the households in the city contain between 2.16 and 2.44 persons.

   (e) Household size in the city follows the normal curve.

   (f) The 95%-confidence level is about right because household size follows the normal curve.

   [77% got at least half credit for this question]

10. A machine makes sticks of butter whose average weight is 4.0 ounces; the SD of the weights is 0.05 ounces. There is no trend or pattern in the data. There are 4 sticks to a package.

    (a) A package weighs ________ give or take ________ or so.

    (b) A store buys 100 packages. Estimate the chance that they get 100 pounds of butter, to within 2 ounces.

    [91% got at least half credit for this question]

11. As part of a statistics project, Mr. Frank Alpert approached the first 100 students he saw one day on Sproul Plaza at the University of California, Berkeley, and found out the school or college in which they enrolled. His sample included 53 men and 47 women. From Registrar's data, 25,000 students were registered at Berkeley that term, and 67% were male. Was his sampling procedure like taking a simple random sample? *[23% got this right]*

12. A geography test was given to a simple random sample of 250 high school students in a certain large school district. One question involved an outline map of Europe, with the countries identified only by number. The students were asked to pick out Great Britain and France. As it turned out, 65.8% could find France, compared to 70.2% for Great Britain. Is the difference statistically significant? Or can this be determined from the information given? *[29% got this right]*

13. One study of grand juries in Alameda County, California, compared the demographic characteristics of jurors with the general population, to see if the jury panels were representative. Here are the results for age. (Only persons 21 and over are considered; the county age distribution is known from Public Health Department data.)

Age	County-wide percentage	Number of jurors
21 to 40	42	5
41 to 50	23	9
51 to 60	16	19
61 and up	19	33
Total	100	66

    Were these 66 jurors selected at random from the population of Alameda County (age 21 and up)?
    [63% got this right]

14. R.E. Just and W.S. Chern claimed that the buyers of California canning tomatoes exercised market power to fix prices. As proof, the investigators estimated the price elasticity of demand for tomatoes in two periods--before and after the introduction of mechanical harvesters. (An elasticity of -5, for instance, means that a 1% increase in prices causes a 5% drop in demand.) They put standard errors on the estimates.

    In a competitive market, the harvester should make no difference in demand elasticity; it only affects supply. However, the difference between the two estimated elasticities--pre-harvester and post-harvester--was statistically significant ($z \approx 1.56$, $P \approx 5.9\%$, one-sided). The investigators tried several ways of estimating the price elasticity before settling on the final version. Comment briefly on the use of statistical tests.
    [69% got this right]

Statistics 2
Fall 1990

Mr. Purves

Final Exam

This three-hour test was taken by 209 students. The average score was 56/100, and the SD was 22.

Print your name ______________________________

Sign your name ______________________________

TA's name ______________________________

Section time ______________________________

To get full credit, you must give reasons and/or show work.

1. Here is a passage from Dr. Dean Edell's column in the *San Francisco Chronicle* of August 1, 1990.

DEMAND AN EXPERIENCED SURGEON

> The more experienced a doctor is, the better. As obvious as that sounds, there are still too many people out there who never ask their surgeons for a history of their work. The importance of knowing is illustrated by this study.
>
> Peter Starck, a surgeon at the University of North Carolina, reviewed 460 heart valve replacement operations and found that only 4 percent of the patients of the three most senior surgeons died. But one junior surgeon lost almost a third of his patients. Since that surgeon was technically the best in the group, says Starek, something was obviously lacking--perhaps the kind of good judgment that grows out of experience...

The last sentence of the paragraph contains the claim that the junior surgeon was obviously lacking something--"perhaps the kind of good judgment that grows out of experience." Is the claim justified by Dr. Starek's evidence? Discuss briefly.

2. According to an article in the November 28, 1990 edition edition of the *San Francisco Chronicle*, 74 percent of the freshman class at UC Berkeley scored over 500 points on the verbal section of the SAT. If the verbal SAT scores for the entire class have an SD of 80 points and follow the normal curve, what is the average?

3. For the 988 men age 18–24 in the HANES sample,

   average height ≈ 70 inches   SD ≈ 3 inches
   average weight ≈ 162 pounds   SD ≈ 30 pounds
   correlation ≈ 0.47

   One man in the sample was 66" tall and weighed 140 pounds. In comparison with the other men in the sample of the same height, this man would be

   a little light   a little heavy.

   Circle one option and explain your choice.

4. A computer printout shows the following descriptive statistics on the relationship between blood pressure and height for a large representative sample of American men:

   average height ≈ 70 inches   average blood pressure ≈ 124 mm
   correlation between height and blood pressure ≈ -0.2

   It also shows the regression equation for predicting blood pressure from height:

   predicted blood pressure = (-0.9 mm per inch) × height + 163 mm

   Is there anything wrong? To answer, choose one option below and then explain your choice.

   (a) There may be something wrong, but there is not enough information here to decide.

   (b) Something is definitely wrong.

5. Two draws are made at random with replacement from the box

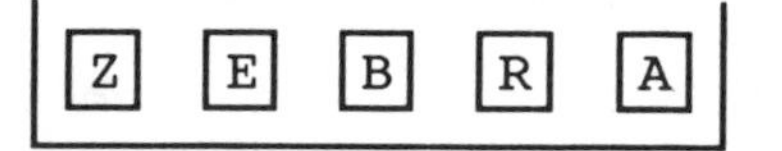

   (a) What is the chance that the letters are different?

   (b) What is the chance of getting a vowel at least once in the two draws?

6. A box of tickets averages out to 75, and the SD is 10. Twenty-five draws are made at random with replacement from this box.

   (a) Find the chance (approximately) that the average of the draws will be in the range 65 to 85.

   (b) Repeat, for the range 74 to 76.

7. A sociologist draws a simple random sample of 500 students from the undergraduates at a large state university. The students are interviewed about their plans upon graduation, and their responses recorded on interview forms. The names of the students, along with the responses, are entered in a computer file in the same order as they were originally chosen for the sample. Unfortunately, before anyone could look at the results, a programming error led to the loss of data for the last 100 students in the file. Eventually, the lost data will be re-entered into the computer from the interview forms, but the sociologist wants to start working on the available data immediately. For example, he finds that out of the 400 students whose responses are still intact, 80 plan to go on to graduate school. Using this 80 out of 400, the sociologist intends to calculate a 95% confidence interval for the percentage of undergraduates at the university who plan to go to graduate school.

   Is this appropriate? If it is, explain why and find the confidence interval; if not, explain why not.

   (Note: Please choose one and only one of the two options. For example do not write, "I don't think the calculation is appropriate, but in case it is, here is how to find the confidence interval" and then go on to do the calculation of the confidence interval.)

8. A simple random sample of 1,000 persons is taken to estimate the percentage of Democrats in a large population. It turns out that 543 of the people in the sample are Democrats. The sample percentage is (543/1,000) × 100% = 54.3%. The SE for the sample percentage of Democrats is figured as 1.6%. True or false, and explain:

   (a) 54.3% ± 3.2% is a 95%-confidence interval for the percentage of Democrats in the population.

   [Parts (b) and (c) are on the next page.]

8. (continued)

(b) 54.3% ± 3.2% is a 95%-confidence interval for the percentage of Democrats in the sample.

(c) There are about two chances in three for the percentage of Democrats in the population to be in the range 54.3% ± 1.6%.

9. The students in a high school physics class made 25 measurements of the weight of a piece of metal about the size of a nickel. On their first weighing, they got 5.29 grams. On their next one, they got 5.36 grams. On their final weighing--the 25th one--they got 5.37 grams. All the results are shown below.

Meas No	Result	Meas No	Result	Meas No	Result	Meas No	Result	Meas No	Result
1	5.29	6	5.47	11	5.58	16	5.59	21	5.49
2	5.36	7	5.49	12	5.64	17	5.56	22	5.48
3	5.37	8	5.50	13	5.72	18	5.53	23	5.47
4	5.41	9	5.52	14	5.65	19	5.52	24	5.39
5	5.44	10	5.54	15	5.62	20	5.50	25	5.37

If it is reasonable to do so, find an approximate 95%-confidence interval for the weight of the piece of metal. If it is not reasonable, explain why not. (The 25 numbers in the table have an average of 5.5 grams and an SD of 0.1 grams.)

10. Here is a quotation from the *San Francisco Chronicle* of June 23, 1990. (The quote has been edited slightly.)

> The same poll contained equal helpings of good news and bad news for Dianne Feinstein's bid to become California's first woman governor.
>
> The survey of 1330 registered voters found Democrat Feinstein with a statistically ________________ lead over Republican Pete Wilson of 52 percent to 48 percent...

One word has been left out. That word is either:

significant or insignificant

For the questions below, you may assume the survey consisted of a simple random sample of 1330 registered California voters.

[The question continues on the next page.]

10. (continued)

    (a) Formulate the null hypothesis implicit in the passage. Translate it into a statement about a box model.

    (b) Calculate the appropriate test statistic and find the observed significance level.

    (c) Was the missing word "significant" or "insignificant."

11. Freshmen at public universities work 12.2 hours a week for pay, on average, and the SD is 10.5 hours; at private universities, the average is 9.2 hours and the SD is 9.9 hours. Assume these data are based on two independent simple random samples, each of size 1,000. Is the difference between the averages due to chance?

    (a) Formulate the null hypothesis as a statement about a box model.

    (b) Repeat (a) for alternative hypothesis.

    (c) Calculate the appropriate test statistic.

    (d) What do you conclude?

SUPPLEMENTARY DATA SETS VERSION 1.0

by

Richard Cutler

Department of Mathematics and Statistics
Utah State University

OVERVIEW

The data sets documented in this appendix are intended to accompany the second edition of *Statistics*. There are two groups of data sets; both are available to users of *Statistics* from the publisher, W. W. Norton & Company, Inc. One group comes on a single 5.25 inch, 360K diskette for IBM compatible personal computers. The other group contains some additional data, and is available on a 3.5 inch, 1.44M diskette for IBM compatible PCs, and on an Apple Macintosh 1.44 M diskette.

All the data sets--with one exception--were derived from the March 1989 Current Population Survey (CPS). To begin with, one large data set (*persons1.dat*) was extracted from the CPS file. The remaining data sets are subsets of *persons1.dat*. (The exceptional data set *scores.dat* consists of quiz, midterm, and final exam scores for a statistics class taught at Utah State University in the Fall Quarter of 1990.)

For every data set there is a corresponding documentation file which explains the origin of the data, the variables in the data set, and the meaning of the different codes. There is also a paragraph in each documentation file explaining which topics the data may be used to illustrate. Appropriate sections of *Statistics* are cited, as well as similar exercises or examples.

The table at the top of the next page contains a complete listing of all the data sets and their documentation files. Starred files are available only on 3.5 inch diskettes; doubly-starred files, only on the 5.25 inch diskettes.

On each diskette, there is an *overview.doc* file, most of which is reprinted here. There are also three files, *input.sas, input.mtb*, and *input.mys*, which contain sample commands for reading the data sets into the statistical software programs SAS, MINITAB, and MYSTAT, respectively.

Data Filename	Documentation Filename
*persons1.dat**	*persons1.doc**
*persons2.dat***	*persons2.doc***
*spouses1.dat**	*spouses1.doc**
*spouses2.dat***	*spouses2.doc***
reg1spou.dat	*reg1spou.doc*
*reg2spou.dat**	*reg2spou.doc**
*reg3spou.dat**	*reg3spou.doc**
*reg4spou.dat**	*reg4spou.doc**
*capeople.dat**	*capeople.doc**
*flpeople.dat**	*flpeople.doc**
*ilpeople.dat**	*ilpeople.doc**
*nypeople.dat**	*nypeople.doc**
txpeople.dat	*txpeople.doc*
div8kinc.dat	*div8kinc.doc*
caincedl.dat	*caincedl.doc*
*flincedl.dat**	*flincedl.doc**
scores.dat	*scores.doc*

* 3.5 inch diskette
** 5.25 inch diskette

EXTRACTING THE SUBSAMPLES

Every month the Bureau of the Census conducts the Current Population Survey for the Bureau of Labor Statistics. In March, additional information is obtained on employment status, occupation, and income for the preceding calendar year. The multistage sampling scheme of the CPS is described in chapter 22 of *Statistics*. The CPS is sampling households; attached to each household is a "sample weight," which is just the inverse of the probability of selection of the household in the CPS.

We began by obtaining a public-use microdata tape for the March 1989 CPS. To form the main data set (*persons1.dat*), we drew 6662 households at random from the tape, with probabilities proportional to their CPS sample weights. That gave us our sample. Because we drew with probabilities proportional to the CPS weights, when we estimate population characteristics from our sample, all households get equal weight; so do all persons in those households. (In the jargon of the trade, this sample is "self-weighting.") Each of our sample individuals represents 14,043 individuals in the U.S. (This is not quite a simple random sample of the U.S., however; see section 22.5 of *Statistics*.)

An example may help. In our sample, there were 1555 persons age 25–29, of whom 385 had college degrees. In the whole U.S., we estimate there were 14,043 × 1555 ≈ 21.8 million persons age 25–29. Among persons in this age group, we estimate that 14,043 × 385 ≈ 5.4 million had college degrees. We can also estimate that, among persons age 25–29 in the U.S., 385/1555 ≈ 25% have college degrees. No weights are needed in this last calculation--the sample is self-weighting.

For our sample, selected information on all persons age 18 and over was entered in the file *persons1.dat*. Variables include age, sex, education, race, marital status, employment status, occupation, personal and family income, and state of residence. Code numbers are reported, so that members of the same family or household can be identified. We also drew a second sample, as a subsample of our first sample; it too is self-weighting. Every individual in the smaller sample represents 56,172 individuals in the whole U.S. Data for the smaller sample are in *persons2.dat*.

In the first sample, 3770 husband-wife families were identified. Selected information on these families was placed in the file *spouses1.dat*. In the second sample, 974 husband-wife families were identified and information on them placed in the file *spouses2.dat*. All the remaining data sets (except *scores.dat*) were obtained from *persons2.dat* or *spouses2.dat* by selecting specific states or regions of residence, or by restricting the age of the members, or by selecting subsets of the variables. These data sets are all self-weighting too.

Any errors in this data processing are our responsibility.

POSSIBLE APPLICATIONS

Descriptive statistics can be computed for these data sets; histograms and scatter diagrams can be plotted. The data can also be used to illustrate inferential procedures in *Statistics*, with a disclaimer of the form: "The following data come from the March 1989 CPS. Assuming the data are a simple random sample from such-and-such a population...." Now, the CPS is not a simple random sample; strictly speaking, the formulas in *Statistics* cannot be used to compute standard errors for CPS sample quantities. (A discussion of how standard errors may be estimated using the half-sample method is given in section 22.5 of *Statistics*.) However, the subsamples are not too much different from simple random samples, so the formulas will give good approximations to the SEs.

persons1.dat, persons2.dat, spouses1.dat, and *spouses2.dat* are large files. They are meant primarily as a resource for instructors. Smaller data sets can be obtained by selecting subsets of records.

reg1spou.dat, reg2spou.dat, reg3spou.dat, and *reg4spou.dat* contain data on husband-wife families in different regions of the U.S. These data sets are helpful for illustrating correlation and regression.

capeople.dat, flpeople.dat, ilpeople.dat, nypeople.dat, and *txpeople.dat* contain data on men and women age 18 and over who live in California, Florida, Illinois, New York, and Texas, respectively. These data sets may be used to illustrate various kinds of hypothesis tests, including the chi-square test for independence and two-sample z-tests. The data may also be used to demonstrate histograms.

caincedl.dat and *flincedl.dat* contain data on income and education for men and women age 25–54, who live in California and Florida, respectively. *div8kinc.dat* contains data on family size and family income for families who reside in the Mountain division of the U.S. These data sets may be used to demonstrate summary statistics, histograms, correlation, and regression.

scores.dat contains quiz, midterm, and final scores for 172 students. This data set may be used to demonstrate summary statistics, histograms, correlation, and regression.

FILE LAYOUT

All data files are flat ASCII text files, with one line per record. (Lines terminate with carriage return and line feed, so files can be manipulated by a text editor.) In *spouses1.dat* and *spouses2.dat*, there is one record per family; otherwise, each record corresponds to a person. Each variable is represented by a string of fixed length, consisting of some number of spaces (ASCII 32) followed by some number of digits. For example, in *persons1.dat*, the 1st variable "STATE" is a two-digit string in "P1–2," i.e., positions 1–2; 11 means Maine, and 95 is Hawaii, according to the table on the next page. The variable "PINCOME" is personal income, in P31–35. An income of $52,615 would be represented as 52615, with one initial space. An income of $199,998--the largest in the sample--would be represented as 199998, with no initial spaces. Some incomes are negative; the left-most digit is preceded by a minus sign (ASCII 45).

persons1.dat

persons1.dat has 12,669 person records, and each record has 19 variables. The record length is 61 characters (spaces count, but not the carriage return and line feed that end the record). The record layout is documented in the table below.

Name and Position	Variable Description	File: *persons1.dat*
STATE P1–2 VARIABLE 1	State of residence, grouped by 4 regions and 9 divisions	

NORTHEAST REGION (REGION 1)

New England Division (Division 1)

Codes: 11 Maine
12 New Hampshire
13 Vermont
14 Massachusetts
15 Rhode Island
16 Connecticut

Middle Atlantic Division (Division 2)

Codes: 21 New York
22 New Jersey
23 Pennsylvania

MIDWEST REGION (REGION 2)

East North Central Division (Division 3)

Codes: 31 Ohio
32 Indiana
33 Illinois
34 Michigan
35 Wisconsin

West North Central Division (Division 4)

Codes: 41 Minnesota
42 Iowa
43 Missouri
44 North Dakota
45 South Dakota
46 Nebraska
47 Kansas

Name and Position	Variable Description	File: *persons1.dat*

STATE
P1–2
VARIABLE 1

State of residence, grouped by 4 regions and 9 divisions, continued

SOUTH REGION (REGION 3)

South Atlantic Division (Division 5)

Codes: 51 Delaware
52 Maryland
53 District of Columbia
54 Virginia
55 West Virginia
56 North Carolina
57 South Carolina
58 Georgia
59 Florida

East South Central Division (Division 6)

Codes: 61 Kentucky
62 Tennessee
63 Alabama
64 Mississippi

West South Central Division (Division 7)

Codes: 71 Arkansas
72 Louisiana
73 Oklahoma
74 Texas

WEST REGION (REGION 4)

Mountain Division (Division 8)

Codes: 81 Montana
82 Idaho
83 Wyoming
84 Colorado
85 New Mexico
86 Arizona
87 Utah
88 Nevada

Pacific Division (Division 9)

Codes: 91 Washington
92 Oregon
93 California
94 Alaska
95 Hawaii

Name and Position	Variable Description	File: *persons1.dat*
SEX P4 VARIABLE 2	Codes: 1 Male 2 Female	
AGE P6-7 VARIABLE 3	Codes: 18-89 Age in years 90 90 years of age or older	
RACE P9 VARIABLE 4	Codes: 1 White 2 Black 3 American Indian, Eskimo or Aleut 4 Asian or Pacific Islander 5 Other	
ETHNICITY P11 VARIABLE 5	Codes: 1 Mexican American 2 Puerto Rican 3 Cuban 4 Central or South American 5 Other Hispanic 6 All other (non-Hispanic) 7 Did not respond	
EDUCATION P13-14 VARIABLE 6	Highest grade of school completed Codes: 0 Did not complete first grade 1 First grade 2 Second grade 3 Third grade 4 Fourth grade 5 Fifth grade 6 Sixth grade 7 Seventh grade 8 Eighth grade 9 Ninth grade 10 Tenth grade 11 Eleventh grade 12 Twelfth grade 13 One year of college 14 Two years of college 15 Three years of college 16 Four years of college 17 Five years of college 18 Six or more years of college	

Name and Position	Variable Description	File: *persons1.dat*
MARITAL P16 VARIABLE 7	Marital status Codes: 1 Married 2 Widowed 3 Divorced 4 Separated 5 Never Married	
NUMKIDS P18 VARIABLE 8	Number of own, unmarried children under 18 years of age Codes: 0-8 Number of children 9 9 children or more	
NUMPERS P20-21 VARIABLE 9	Number of persons in family Codes: 1-39 Number of persons	
EMPSTAT P23 VARIABLE 10	Employment status Codes: 0 Armed forces Civilian labor force 1 Working 2 With a job, not at work (e.g., on vacation) 3 Unemployed Not in labor force 4 Keeping house 5 Attending school 6 Unable to work 7 Retired 8 Other	
FULLPART P25 VARIABLE 11	Part/Full-time worker status Codes: 0 Armed forces 1 Not in labor force 2 Full-time 3 Part-time 4 Unemployed	

Name and Position	Variable Description File: *persons1.dat*
CLASSWORK P27 VARIABLE 12	Class of worker Codes: 0 Not in civilian labor force Civilian labor force (Current or most recent job was...) 1 Private sector 2 Federal government 3 State government 4 Local government 5 Self-employed (incorporated) 6 Self-employed (not incorporated) 7 Without pay (e.g., in a family business) 8 Unemployed, no previous experience
INDUSTRY P29-30 VARIABLE 13	Describes current or most recent job, for persons in the civilian labor force Codes: 0 Not in civilian labor force *CIVILIAN LABOR FORCE* 1 Agriculture 2 Mining 3 Construction Manufacturing 4 Manufacturing--durable goods 5 Manufacturing--nondurable goods Transportation, communications, and other public utilities 6 Transportation 7 Communications 8 Utilities and sanitary services Wholesale and retail trade 9 Wholesale trade 10 Retail trade 11 Finance, insurance, and real estate

Name and Position	Variable Description	File: *persons1.dat*

INDUSTRY
P29-30
VARIABLE 13

Continued

Services

12 Private household
13 Business and repair
14 Personal services, except household
15 Entertainment
16 Hospital
17 Medical, except hospital
18 Educational
19 Social services
20 Other professional services

Other

21 Forestry and fisheries
22 Public administration

Unemployed

23 Unemployed, was in armed forces
24 Unemployed, no previous experience

OCCUPATION
P32-33
VARIABLE 14

Describes current or most recent job, for persons in the civilian labor force

Codes: 0 Not in civilian labor force

CIVILIAN LABOR FORCE

Codes: Managerial and professional

1 Executive, administrative, and managerial
2 Professional specialty

Technical, sales, and administrative support

3 Technicians and related support
4 Sales
5 Administrative support, including clerical

Name and Position	Variable Description	File: *persons1.dat*
OCCUPATION P32-33 VARIABLE 14	Continued Service 6 Private household 7 Protective service 8 Other service Operators, fabricators, and laborers 9 Precision production, craft, and repair 10 Machine operators, assemblers, and inspectors 11 Transportation and material moving 12 Handlers, equipment cleaners, etc. 13 Farming, forestry, and fishing Unemployed 14 Unemployed, was in armed forces 15 Unemployed, no previous experience	
PINCOME P35-40 VARIABLE 15	Personal income in 1988 Codes: Dollar amount of income, positive or negative	
FINCOME P42-47 VARIABLE 16	Family income in 1988 Codes: Dollar amount of income, positive or negative	
FAMCODE P49-55 VARIABLE 17	A unique identifier for each family Codes: 1-9999999	
PERSCODE P57-58 VARIABLE 18	Person's sequence number Codes: 1-39 Person's sequence number in household	
SPOUCODE P60-61 VARIABLE 19	Code for spouse's sequence number Codes: 0 No spouse 1-39 Spouse's sequence number in household	

SAMPLE RECORDS

The first record in *persons1.dat* is shown below; certain positions in the record are marked for reference.

```
Position| 2 4      9    14   18 21   25    30 33       40      47        55 58
        | | |      |     |    |  |    |     |  |        |       |         |  |
  Record|11 1 32 1 6 16 1 1   3 1 2 5 11  1   52615  78230   171701  1
```

This record can be decoded as follows.

STATE, P1–2. Code 11 corresponds to Maine. This person lives in Maine.

SEX, P4. Code 1 means male.

AGE, P6-7. This person was 32 years of age at the time of the survey.

RACE, P9. Code 1 means the person is white.

ETHNICITY, P11. Code 6 means the person is not Hispanic.

EDUCATION, P13-14. Code 16 means the person had completed 16 years of school (i.e., a 4 year college degree).

MARITAL, P16. Code 1 means the person was married.

NUMKIDS, P18. This person has one unmarried child under the age of 18.

NUMPERS, P20-21. There are 3 people in this person's family: this person, his wife (we already know the person is male and married), and that unmarried child under the age of 18.

EMPSTAT, P23. Code 1 means the person was working.

FULLPART, P25. Code 2 means the person worked full-time.

CLASSWORK, P27. Code 5 means the person was self-employed (incorporated).

INDUSTRY, P29-30. Code 11 is for finance, insurance, and real estate.

OCCUPATION, P32-33. Code 1 is for executive, administrative, and managerial jobs.

(The first record is repeated below.)

PINCOME, P35-40. This person earned $52,615 in 1988.

FINCOME, P42-47. The 1988 family income for this person was $78,230.

FAMCODE, P49-55. The family code for this person is 171701. This code may be used to identify other members of the same family.

PERSCODE, P57-58. This person's sequence number is 1.

SPOUCODE, P60-61. The sequence number of this person's spouse is 2.

The first 9 records in *persons1.dat* are shown below:

Position	2	4		9		14		18	21		25		30	33	40	47	55	58	61
Record 1	11	1	32	1	6	16	1	1	3	1	2	5	11	1	52615	78230	171701	1	2
Record 2	11	2	31	1	6	16	1	1	3	1	2	4	18	2	25615	78230	171701	2	1
Record 3	11	2	64	1	6	12	3	0	1	4	1	0	0	0	3966	3966	4596701	1	0
Record 4	11	1	46	1	6	6	3	0	1	8	1	0	0	0	4362	4362	4596702	2	0
Record 5	11	2	32	1	6	10	3	0	1	1	3	1	10	4	2000	2000	4596601	1	0
Record 6	11	1	32	1	6	12	1	2	4	1	2	6	3	9	36010	44520	5466701	1	2
Record 7	11	2	32	1	6	13	1	2	4	1	2	4	18	2	8510	44520	5466701	2	1
Record 8	11	2	55	1	6	13	4	0	2	1	2	6	14	8	13000	28008	5465801	1	0
Record 9	11	2	32	1	6	13	5	0	2	8	1	0	0	0	15008	28008	5465801	4	0

Usually, a household consists of a family. (Sometimes, there are several families living together, and perhaps unrelated individuals too; about 25% of households consist of single individuals living alone.) If a household comes into the sample for *persons1.dat*, all persons age 18 and over are listed in the file. As you can see from the display, the first two records belong to one family: FAMCODE, P49–55, has the same value for each record. (Being under 18, the child in the family was excluded from *persons1.dat*.)

The 2nd record in *persons1.dat* is for the spouse of the person in record 1. Both records have the same family code, and the person sequence number in the 2nd record is the same as the spouse sequence number in the 1st record. (Person sequence number is the variable PERSCODE, P57–58; and spouse sequence number is the variable SPOUCODE, P60–61.)

In FAMCODE, P49–53 identifies the household; P54–55 identifies families (or unrelated individuals) within the household. Records 3 and 4 belong to two unrelated individuals living together; the person sequence number (PERSCODE, P57–58) is within household. Record 5 belongs to an individual in a one-person household. Records 8–9 belong to a family, probably a mother and her daughter. The person sequence number skips from 1 to 4: this is not a transcription error.

Notes: EMPSTAT and FULLPART code status during the week before the interview. (The labor force consists of employed plus unemployed persons; unemployed persons are available for work and looking for work.) If the respondent was employed, then CLASSWORK, INDUSTRY, and OCCUPATION code the current job; if the respondent was unemployed, these variables code the most recent job. Unemployed persons with no previous job (i.e., new entrants to the labor force) are coded 8 on CLASSWORK, with INDUSTRY=24 and OCCUPATION=15. Persons who were not in the civilian labor force are coded 0 on CLASSWORK, INDUSTRY, and OCCUPATION.

persons2.dat

persons2.dat is a subsample of *persons1.dat*; the record layout, variables, and codes are defined as for *persons1.dat*.

spouses1.dat

"Husband-wife" families in *persons1.dat* consist of a husband, a wife, and perhaps some children. (Other families comprise a woman and some children; and there are unrelated individuals in some households.) *spouses1.dat* has one record for each of the 3770 husband-wife families in *persons1.dat*, and extracts data on age, educational level, and personal income for the husband and wife, as well as family variables, such as family size, family income, and number of children under 18 years of age.

spouses1.dat and data sets derived from it (e.g., by taking only families resident in the Northeast) may be useful for illustrating correlation and regression (chapters 8–12 of *Statistics*). This is because most of the variables in *spouses1.dat* are paired; e.g., wife's income and husband's income, wife's educational level and husband's educational level.

spouses1.dat contains 3770 records and each record has 10 variables. The record length is 40 characters (spaces count, but not the carriage return and line feed that end the record). The record layout is documented in the table below. The codes for the variables are as defined for *persons1.dat*.

If there are members of the family other than the husband and wife, the family income may differ from the sum of the personal incomes of the husband and wife.

Some exercises and examples in *Statistics* which involve variables like those in *spouses1.dat*: Example 3 in section 5 of chapter 10 and review exercise 1 in chapter 8 concern IQs of husbands and wives. Review exercise 3 in chapter 10 concerns educational levels of husbands and wives. The lead example in section 1 of chapter 12 discusses the use of the regression line to summarize the relationship between income and education. Exercises 1 and 3 in set A of chapter 12 also concern income and education.

Name and Position	Variable Description File: *spouses1.dat*
STATE P1–2 VARIABLE 1	State of residence, grouped by 4 regions and 9 divisions
NUMKIDS P4 VARIABLE 2	Number of own, unmarried children under 18 years of age
NUMPERS P6–7 VARIABLE 3	Number of persons in family
AGEWIFE P9–10 VARIABLE 4	Age of wife in years
AGEHUSB P12–13 VARIABLE 5	Age of husband in years

Name and Position	Variable Description	File: *spouses1.dat*
EDUWIFE P15–16 VARIABLE 6	Highest grade of school completed by wife	
EDUHUSB P18–19 VARIABLE 7	Highest grade of school completed by husband	
PINCWIFE P21–26 VARIABLE 8	Wife's personal income in 1988	
PINCHUSB P28–33 VARIABLE 9	Husband's personal income in 1988	
FINCOME P35–40 VARIABLE 10	Family income in 1988	

spouses2.dat

spouses2.dat was derived from the second, smaller sample of households in exactly the same way that *spouses1.dat* was obtained from the larger sample. This file has 974 records. The record layout is exactly the same as for *spouses1.dat*.

reg1spou.dat, reg2spou.dat, reg3spou.dat, reg4spou.dat

reg1spou.dat, reg2spou.dat, reg3spou.dat, and *reg4spou.dat* were obtained from *spouses2.dat* by restricting the states of residence to the Northeast, Midwest, South, and West regions of the U.S., respectively. *reg1spou.dat* contains 193 records, *reg2spou.dat* contains 231 records, *reg3spou.dat* contains 364 records, and *reg4.dat* contains 186 records. The record layout is exactly the same as for *spouses1.dat*.

capeople.dat, flpeople.dat, ilpeople.dat, nypeople.dat, txpeople.dat

These five data sets were obtained from *persons2.dat* by restricting the states of residence of the persons to California, Florida, Illinois, New York, and Texas, respectively, and by selecting only 8 variables.

capeople.dat contains 411 records, *flpeople.dat* contains 179 records, *ilpeople.dat* contains 141 records, *nypeople.dat* contains 261 records and *txpeople.dat* contains 243 records. The record layout for all of these files is the same and is documented in the table below. The codes for these variables are as defined for *persons1.dat*.

These data sets may be used to illustrate certain kinds of hypothesis tests--in particular, the chi-square test for independence (section 28.2 of *Statistics*) and two-sample z-tests for means and proportions (section 27.2). Here are some exercises in *Statistics* which use similar data. Exercise 2 in set E of chapter 28 concerns a cross-tab of sex and marital status. Review exercise 2 in chapter 28 concerns a cross-tab of marital status and employment status. Review exercise 3 in chapter 27 involves two-sample z-tests for means of incomes and proportions of persons employed.

Name and Position	Variable Description	File: *??people.dat*
SEX P1 VARIABLE 1	Sex	
AGE P3–4 VARIABLE 2	Age in years	
RACE P6 VARIABLE 3	Race	
ETHNICITY P8 VARIABLE 4	Ethnicity	

Name and Position	Variable Description	File: *??people.dat*
MARITAL P10 VARIABLE 5	Marital status	
EMPSTAT P12 VARIABLE 5	Employment status	
FULLPART P14 VARIABLE 6	Part/Full-time worker status	
PINCOME P16–21 VARIABLE 8	Personal income in 1988	

caincedl.dat, flincedl.dat

These data sets were obtained from *persons2.dat* by restricting the states of residence of the persons to California and Florida, respectively; by restricting the ages of the persons to the range 25–54; and by selecting only 3 variables.

caincedl.dat contains 258 records and *flincedl.dat* contains 98 records. The record layout for these files is the same and is documented in the table on the next page. The codes for the variables are as defined for *persons1.dat*.

caincedl.dat and *flincedl.dat* may be useful for demonstrating histograms, correlation, and regression (chapters 3 and 8–12 of *Statistics*). Figures 1–2 in chapter 3 are histograms for income. Figure 5 is a histogram for educational level. The relationship between education and income is discussed in some detail in section 1 of chapter 12; exercises 1 and 3 in set A of chapter 12 are also concerned with income and education data.

Name and Position	Variable Description	File: *??incedl.dat*
SEX P1 VARIABLE 1	Sex	
EDUCATION P3–4 VARIABLE 2	Highest grade of school completed	
PINCOME P6–11 VARIABLE 3	Personal income in 1988	

div8kinc.dat

This data set was obtained from *spouses2.dat* by restricting the states of residence of the families to the Mountain division of the U.S. and by selecting only 2 variables *div8kinc.dat* contains 47 records. The layout of the records is given in the table below. The codes for the variables are as defined for *persons1.dat*.

This data set may be used to illustrate histograms. In chapter 3 of *Statistics*, figures 1–2 are histograms of family incomes, and figure 6 is a histogram of family size.

Name and Position	Variable Description	File: *div8kinc.dat*
NUMPERS P1–2 VARIABLE 1	Number of persons in family	
FINCOME P4–9 VARIABLE 2	Family income in 1988	

scores.dat

This data set contains quiz, midterm, and final exam scores for 172 students in an undergraduate statistics class at Utah State University in the Fall Quarter of 1990. Each student in the class attended lectures on Monday, Wednesday, and Friday; and a small (20–40 students) recitation section on Tuesday and Thursday. Each recitation section was conducted by a different teaching assistant, and students stayed in the same section for the whole quarter. Quizzes were given in the recitations on Thursdays, except for the first and last weeks and the week of the midterm. There was one midterm and a final exam. Variables in *scores.dat* and their codes are given in the table below.

Name and Position	Variable Description	File: *scores.dat*
SECTION P1 VARIABLE 1	Recitation section attended by student Codes: 1–5 Recitation section number	
QUIZ P3–5 VARIABLE 2	Overall quiz scores Codes: 0–180 Sum of six best quiz scores	
MIDTERM P7–9 VARIABLE 3	Midterm score Codes: 0–100 Score on midterm (out of 100)	
FINAL P11–13 VARIABLE 4	Final exam score Codes: 0–200 Score on final (out of 200)	

This data set may be useful for demonstrating histograms, the normal approximation for data (the midterm and final scores are approximately normally distributed), correlation, and regression. Some exercises in *Statistics* which involve similar kinds of data: exercises 3, 4, 5, and 6 in set A of chapter 3; review exercises 1, 2, 4, 5, and 7 in chapter 5; exercise 7 in set A of chapter 8; exercises 1 and 2 in set C of chapter 10.